CONTENTS

Foreword   iii

Problems of Internal Inhibition
U.G. GASSANOV

1. Introduction   3
2. Nature of Internal Inhibition   4
3. Localization of Internal Inhibition   14
4. Mechanisms of Internal Inhibition   20
5. Gnostic Neuronal Network   26

References   30

Index   35

The Dominant and the Conditioned Reflex
R.A. PAVLYGINA

1. Basic Postulates of the Ukhtomskii Dominant Theory   39
2. Methods of Investigation   42
3. Forward and Backward Connections in the Case of the Dominant   44
4. Reciprocal Inhibition in the Case of the Dominant   51
5. Goal-Directed Reactions Developed on the Basis of the Dominant
   Discontinuation   52
6. The Role of the Dominant in the Formation of Conditioned
   Reflexes   61
7. Conclusion   65

References   67

Index   71

FOREWORD

Volume 5 of *Physiology and General Biology Reviews* completes the series of reviews, begun in the fourth volume, on the physiology of the higher nervous activity of man and animals, i.e. on the physiological mechanisms of the brain which underlie behavior and mental activity. This discipline is close, in subject and method of investigation, to comparative and psychological physiology, psychophysiology, neuropsychology, neuroethology, and some others, actively developed in Western countries. The physiology of higher nervous activity is characterized by a constant trend towards integration of neural sciences with psychology, unlike behaviorism and cognitive psychology, which study the behavior and mental activity of man and animals without regard for brain structure and functions.

I.P. Pavlov (1849–1936) is rightfully considered the founder of physiology of higher nervous activity (and the author of this term), although I.M. Sechenov (1829–1905), "father of Russian physiology" and materialistic psychology and author of the famous book *Refleksy Golovnogo Mozga* (*Brain reflexes*) (1863), actually generated the ideas which constituted the basis of this new discipline. Sechenov's ideas were also developed by the physiological school of the University of Petersburg, now Leningrad, where one of Pavlov's students, N.E. Vvedenskii (1855–1922), the author of the doctrine of excitable systems inhibition ("Vvedenskii inhibition") worked. A representative of this school, A.A. Ukhtomskii (1875–1942), put forward a theory of the dominant, a reflex system predominating at a given moment and determining the vector of goal-directed behavior. This school also included I.S. Beritashvili (1884–1974) who later moved to Tbilisi and founded the Georgian school of neurophysiologists. Beritashvili investigated the basic differences between two forms of conditioned reflexes, those which require numerous combinations of the stimulus and reinforcement and those forming very fast, after one reinforcement as a rule. Nowadays these two forms of learning are described in literature as declarative and nondeclarative memory.

Pavlov's students and followers made substantial contributions to the physiology of higher nervous activity. L.A. Orbeli (1882–1958), L.G. Voronin (1908–83) and L.V. Krushinskii (1911–84) developed a comparative and evolutionary physiology of higher nervous activity. P.K. Anokhin (1898–1974) studied self-regulation in a living organism on the basis of correcting feedback (theory of functional systems). These studies belong to the so-called "cybernetic" trend in the investigation of brain activity, although Anokhin's ideas had been formulated long before the era of cybernetics. E.A. Asratyan (1903–81) developed the concept of bilateral conditioned relationships as a reflex basis of motivated behavior and, hence, included the problem of motivation in the framework of Pavlovian heritage. Conditioned reflex regulation of visceral organs and the brain's general functional condition were studied by K.M. Bykov (1886–1959), V.N. Chernigovskii (1907–81) and P.S. Kupalov (1888–1964). E.N. Sokolov is the author of classical studies of physiological mechanisms of the orienting reflex. M.N. Livanov (1907–86) and V.S. Rusinov started pioneering electrophysiological studies of conditioned reflexes in this country.

The authors of these reviews might be considered as students of Pavlov's students. Behavioral, neuronal and neurochemical mechanisms of conditioned reflex formation, learning and memory come first in these authors' studies and in those of scientists cited in the references. At the same time modern experimental methods and computer processing of experimental data are increasingly being used. The conceptual basis of the conditioned reflex theory has also been developed and now includes such phenomena as motivation, emotion, forecasting of adaptive results, etc. The modern physiology of higher nervous activity can no longer be considered a modification of neobehaviorism. Its systemic and synthetic character and constant tendency towards analytical neurophysiology are apparent for all unbiased readers.

The remainder of the volume will be devoted to neuroendocrinological studies.

T.M. TURPAEV
P.V. SIMONOV

Sov. Sci. Rev. F. Physiol. Gen. Biol., Vol. 5, 1991, pp. 1–35
Photocopying permitted by license only
© 1991 Harwood Academic Publishers GmbH
Printed in the United Kingdom

PROBLEMS OF INTERNAL INHIBITION

U.G. GASSANOV

Institute of Higher Nervous Activity and Neurophysiology, USSR Academy of Sciences, Moscow

ABSTRACT

The published data convincingly show that the inhibitory conditioned reflex is directly related in its origin to the positive conditioned reflex, since the appearance of inhibitory behavior requires cancellation of the reinforcement, which was previously regularly applied during elaboration of a certain adaptive reaction. This means that physiological mechanisms of internal inhibition must be closely associated with pre-existing mechanisms of associative learning. This close affinity of two conditioned processes with opposite external (behavioral) manifestations provides enough grounds to suggest some common properties of nervous activity during inhibitory and positive conditioned reflexes. According to the concept developed in this paper, any form of learning, including internal inhibition, has in its neurophysiological basis a gnostic neuronal network, specific in function and local in organization. During internal inhibition the gnostic network appears as if in pure form, whereas, during positive reflexes, it is complicated by additional relations branched in space. In both cases the gnostic network is the central link in conditioning and contains memory traces.

KEYWORDS: Internal inhibition, behavior, slow potentials, EEG, EP, firing rate, neuronal network, conditioning

1

1. INTRODUCTION

It has become a tradition to relate all experimental studies of the nervous mechanisms of internal inhibition to two most important problems, namely (1) the nature of internal inhibition and (2) its localization. These problems have caused much controversy since inhibitory conditioned reflexes were discovered.

First, it is assumed that it is the inhibitory process that suppresses the elaborated reaction. According to the ideas of I.P. Pavlov, inhibition, elaborated on omission of reinforcement, was localized in the cortical projection of the conditioning signal (CS). However, special experiments on conditioned reflexes suggested the possible localization of the inhibitory process in different links of the reflex arc.

Further development of the internal inhibition theory was centered around the question of the nature of this behavioral phenomenon. Researchers tended to believe that inhibitory conditioned reflexes form according to the same physiological patterns as positive conditioned reflexes and that there is no reason to search for inhibitory processes. The question as to where to look for the inhibitory process has gradually transformed into the question — what is to be looked for during internal inhibition?

At the end of the 1940s a new trend appeared in studies of physiological mechanisms of the conditioned reflex, namely behavioral electrophysiology. Physiological analysis of the mental activity of animals according to the external, motor or vegetative reaction gave way to analytic studies of the immediate electrical reactions of neuronal structures during performance.

The first experiments with recording of bioelectrical activity seemed to confirm the classical ideas of Pavlov concerning the inhibitory nature of internal inhibition and the localization of this inhibition in the CS projection. However, as experimental data accumulated, doubts again arose as to the explanation of inhibitory conditioned reflexes in terms of a blockade of nervous activity.

Published data have convincingly shown that the inhibitory conditioned reflex is, in its origin, directly related to positive conditioned reflexes, since the appearance of inhibitory behavior requires cancellation of that reinforcement which was previously regularly applied during elaboration of a certain adaptive reaction. This means that the physiological mechanisms of internal inhibition must be closely associated with already existing mechanisms of associative

3

learning. This close affinity of two conditioned processes with opposite external (behavioral) manifestations provides enough grounds to suggest some common properties of nervous activity during inhibitory and positive conditioned reflexes.

In the present paper, internal inhibition is considered a derivative of associative learning and is used for understanding neuronal mechanisms of learning.

2. NATURE OF INTERNAL INHIBITION

The theoretical ideas of the Pavlovian school concerning internal inhibition were presented in final form by Pavlov in his last papers. The inhibition which develops gradually when the CS "is not accompanied by its unconditioned stimulus" was called "internal, active conditioned inhibition" (Pavlov, 1951, p. 224).

Let us note two important definitions of internal inhibition: elaboration and active pattern. Common features of conditioned excitation and inhibition were described during Pavlov's lifetime in his laboratories. These include the wave-like pattern of these processes, the ability to irradiate and concentrate, trace phenomena, disturbances under the influences of external stimuli and the law of stimulus strength. There were far fewer differences: fragility, easier susceptibility of internal inhibition to external influences and, naturally, reciprocity with reference to excitation.

A sudden, external, moderate stimulus resuppressed the inhibitory reaction and did not affect the positive reflex. If the urgent prolongation of the CS action (so-called continuous extinction) led to salivation being stopped, the same procedure with the inhibitory signal could induce a neurotic condition. These facts suggest different levels of nervous activity during two elaborated reflexes, opposite in effect, with predominance of this level in responses to the inhibitory signal.

All these data were obtained in Pavlov's laboratories and did not raise doubts as to the functional antagonism of the cortical processes. This is quite understandable. There is only one answer to the crucial question as to why the elaborated reflex is inhibited when reinforcement is canceled—because an inhibitory process develops in the cortex. By identifying the nature of the conditioned inhibition with the other types of inhibition of nervous activity, such as external inhibition beyond threshold inhibition and sleep, Pavlov proved that internal inhibition is a true inhibition and arises due to exhaustion of the cortical cells. I will not discuss internal inhibition as an active process, on the one hand, and as exhaustion of the nervous cells, on the other, but will turn to the history of concepts concerning the nature of internal inhibition.

The problem of localization of the inhibitory process led to concepts about the nature of the internal inhibition. Although many researchers believed that inhibition was of cortical origin, its localization in the CS projection was held in doubt even among Pavlov's students. For example, Perel'tsveig in 1907 and Kasherininova in 1909 (Asratyan, 1970) concluded that conditioned inhibition is localized in the nervous structure of the unconditioned reflex. During recent decades it was convincingly shown that inhibition is absent in cortical sites of the CS. The most demonstrative and goal-directed are studies carried out in the laboratory of Asratyan (Asratyan, 1970). On the basis of experiments with conditioned switching and with the extinction of binary (heterogeneous, according to reinforcement, and bieffector reflexes to one stimulus) and single conditioned reflexes, he showed that, during conditioned inhibition, the inhibitory process is absent not only in the cortical projection of the CS but also in the cortical projection of the unconditioned reflex. Further experiments fully confirmed these conclusions. An idea about localization of conditioned inhibition in elements (pathways) of temporary connection was proposed after analysis of the facts obtained. It was suggested that as inhibition intensified, it could spread over corresponding cortical projection zones. While discussing the problem of localization of cortical inhibition, Asratyan touches upon the problem of the mechanisms of this inhibition. He believes that, during reinforcement, the elements of temporary connection are inhibited according to the mechanism of negative induction from the excited structure of the unconditioned reflex. This inductive inhibition prevents the middle link of the conditioned reflex from functional exhaustion. Cancellation of reinforcement promotes long-term excitation of elements of the temporary connection and, as a result, leads to development of conditioned inhibition in them.

According to Kupalov's concept (1955), unconditioned inhibition is considered a basis of conditioned inhibition. Unconditioned inhibition includes both negative induction and superthreshold inhibition. Cancellation of reinforcement triggers unconditioned inhibition.

A new integral reaction during unreinforcement is considered as crucial in internal inhibition (Anokhin, 1968). As a result of competition of two stimuli, biologically positive to the previously reinforced stimulus and biologically negative during unreinforcement, the previously elaborated reflex is inhibited due to repeated omission of reinforcement of the new conditioning. Anokhin negates the localization of conditioned inhibition in the signal analyzer and proposes that inhibition might arise in different nodules of systemic organization of the conditioned reflex.

Hence, two important concepts stem from fundamental studies during the post-Pavlovian period of physiology of higher nervous activity. First, cortical inhibition in response to the unreinforced signal is not localized in the structure of the signal stimulus. This new, experimentally substantiated conclusion is an important contribution to the problem of localization of internal inhibition. Second, cancellation of reinforcement is a biologically important stimulus. Since the latter concept will be crucial to later discussion, it should be stressed that it concerns the cancellation of reinforcement previously associated with the phase and environment stimuli, rather than the absence of reinforcement. It is this circumstance that ascribes physiological directionality and biological sense to omission of reinforcement.

It can be seen that the search for localization and mechanisms of internal inhibition preserved the basis of the theory which visualized internal inhibition as a cortical inhibitory process, although the initial classical concept was changed.

However, further studies of internal inhibition raised doubts as to the inhibitory nature of conditioned inhibition. Konorski (1970) is one of the first and consistent supporters of this trend of research. He considers positive and inhibitory conditioned reflexes antagonistic and substantiates their elaboration by referring to connections between neurons of projection of the positive signals with neurons of the structure of a reinforcing unconditioned agent in the case of positive conditioned reflex, and with neurons of projection of the "absence of unconditioned agent" during the inhibitory reflex. In this concept we encounter a special kind of excitation induced by cancellation of reinforcement. Unlike the previous concept, cancellation of reinforcement is always related here to the stimulation of definite centers (or semicenters) responsible for the antagonistic condition of the structure of confirming unconditioned stimulation (for example, cessation of fear during painful stimulation). It is important that Konorski characterizes conditioned inhibition as an ordinary conditioned reflex, which receives regular reinforcement, antagonistic in physiological effect, to the reinforcement of the positive reflex. He stresses that elaboration of negative (inhibitory) conditioned reflexes does not affect previously formed connections of the positive signal: "...we have no evidence of the formation of inhibitory relationships between two groups of neurons, since all available examples of inhibition can be explained by the formation of excitatory relationships between a group of transmitting neurons and a group of reciprocal receiving neurons. Hence, in these cases 'inhibitory conditioned reflex' is no more and no less inhibitory than any other conditioned reflex" (Konorski, 1970, p. 272).

In these investigations not only the question of localization of internal inhibition is raised, inhibition during elaboration of inhibitory, in effect, conditioned reflexes, is negated.

The idea of internal inhibition without inhibition has been proposed in many studies. In 1946, Skipin described the strengthening of the motor food reaction of dogs against the background of extinction of secretory reaction and concluded that "extinctive inhibition is, in essence, its mechanism, one of the phases of excitation and viz. overexcitation" (Skipin, 1940, p. 462).

Amsel (1973) obtained the so-called effect of frustration during unreinforcement. If a rat running through two chambers got a food reward in both, omission of reinforcement in the first chamber accelerated the run in the second. Experiments with unreinforcement gave an unambiguous result: neither frustration nor internal inhibition are the result of the formation of physiological inhibition, they are excitatory processes. The concept of the biological importance of cancellation of reinforcement was further developed by Belenkov (1985).

The search for an inhibitory process produced two fundamental results. First, cancellation of reinforcement is a stimulus of new reinforcement and, second, during internal inhibition, expressed in the inhibition of the previously elaborated effector reflex, nervous activity is activated in the same way as in response to the positive signal. It was shown that, during elaboration of the inhibitory reflex, even the excitability in the analyzer of signal stimulus increased. It was found in experiments on dogs (Pressman and Varga, 1960) that the sensitivity of nervous structures perceiving inhibitory CS undergo changes. In these experiments, a motor reflex to an air puff towards the eye was elaborated and, after the reflex was strengthened, extinction followed. As the motor reaction weakened, the eye reaction to the air movement strengthened.

In estimates of the level of excitability, direct studies of the functional state of the central structure of the coupled stimuli are more demonstrative. Chilingaryan and Romanovskaya (see Asratyan, 1970) tested the cortical zone of a paw in dogs with electric current, while combining two variants of conditioning. In one variant, a distant stimulus was combined with electrical stimulation of the motor cortex and in another with the paw stimulation. After the motor conditioned reflex was elaborated, it was extinguished. As extinctive inhibition developed, the excitability of cortical cells first increased and then returned to the initial level and fell only at the stage of complete extinction. This result not only indicates the stage of increased excitability, it also attracts attention to the necessity of differentiation of extinction by duration.

The many applications of electrophysiological methods to the elucidation of the mechanisms of learning did not live up to the hopes initially pinned on them. The experimental surge of the 1950s and 1960s was replaced by a much more moderate, better thought out combination of electrophysiological and behavioral studies. Electrophysiology of the conditioned reflex confronted the same problems that had been raised before studies of bioelectrical reactions started. In any case, this is true of internal inhibition. As in classical studies of conditioned reflex, many results were published concerning the identity of sleep and internal inhibition and the inhibitory condition of cortical cells during elaboration of inhibitory reflexes. Slow synchronous fluctuations of potentials recorded both during sleep and internal inhibition served as the basic index of inhibition.

Anokhin (1968) has discussed this problem in detail, and critically. At present, researchers define inhibitory states less strictly according to EEG. The published data show that biorhythms can be synchronized and desynchronized during elaboration of both positive and inhibitory conditioned reflexes (Kotlyar, 1963; Kratin, 1967; Weinberger et al., 1967; Anokhin, 1968; Livanov, 1972). This leads to the same conclusion which was drawn on the basis of behavioral experiments, namely positive and inhibitory conditioned reflexes do not differ from each other in the parameters of activity of nervous structures.

In studies of evoked potentials, for example EEG studies, analysis of the results was often based on ideas of blocking the inhibitory process during behavioral inhibition. Theoretical treatment of the electrophysiological data repeated the development of the theory of internal inhibition according to the effector index. This was to a great extent enhanced by the dominating neurophysiological concepts concerning the activatory and inhibitory brain systems, inhibitory origin of certain forms of potentials and hyperpolarization of nerve cells.

Each of these discoveries survived the peak of its influence on the development of behavioral physiology. Numerous experimental studies, especially in the field of learning and memory, have shown that all neurophysiological theories applied to learning frequently raise new problems and promote solutions of the existing questions of integrative activity of a learning brain. In addition, it is the electrophysiological study of the mechanisms of learning that established the limits of application of general neurophysiological concepts.

For example, amplitude changes in evoked potentials were observed in very diverse manifestations during elaboration of positive conditioned reflexes. Comparison of these fluctuations with the threshold of induced response has shown that the evoked response amplitude can be increased or decreased against a background of high excitability of the cells which generate these fluctuations (Gassanov, 1972). In addition, comparison of different phases of the evoked potential with the spike activity has not led to definite conclusions. For example, in the auditory cortex the primary positivity only coincides with the firing rate of a neuron (Reskorla, 1973).

It seems evident that positive reflexes are due to activation of excitatory synapses, whereas the inhibitory ones are due to inhibitory synapses. However, experiments to record effector reactions to an inhibitory signal made this point of view invalid as well (Konorski, 1970, 1973).

A hyperpolarization hypothesis of internal inhibition based on the hyperpolarization theory of nervous cell inhibition (Eccles, 1966) has also appeared (Supin, 1969; Shul'gina, 1978). The idea of the inhibitory nature of internal inhibition was developed most consistently by Shul'gina (1978, 1984). She proved that all kinds of internal inhibition, namely extinction of orienting and defense reflexes, natural and narcotic sleep, are based on hyperpolarization in the form of slow fluctuations of potentials or inhibitory pauses in the background and evoked activity of neurons.

Extinction of the orienting reaction, which is called negative learning or habituation, is widely used as a model of the inhibitory reflex. Habituation is characterized by obligatory suppression of all studied reactions irrespective of whether they are referred to behavior, activity of the vegetative nervous system, bioelectrical activity or intracellular potentials. Habituation is a specific process since changes in the parameters of a signal, in response to which extinction is produced, recover the extinguished reaction (Sokolov, 1970).

While considering habituation as an analog of internal inhibition, some researchers extend the mechanisms of negative learning to cover the formation of various kinds of internal inhibition.

Seemingly, this approach to studying internal inhibition is justified, since, during latent inhibition (habituation), elaboration of a positive conditioned reflex is complicated. However, experimental data were obtained which suggest that latent inhibition complicates elaboration of both positive and inhibitory conditioned reflexes (Reskorla, 1973).

The results of studies of the evoked potentials are very diverse. This appears to be explained both by the configuration complexity of the induced potentials in wakeful animals, especially the potentials of cortical origin, and the method of study. I would like to dwell here on studies which provided the results closely related to the problem of the nature of internal inhibition.

Van Tai-an and Nezlina (1963) elaborated a conditioned reflex to a combination of light and electroshock in monkeys. The applied light flashes did not induce a primary response in the parietal and visual cortex, but flashes of subthreshold intensity started to induce cortical responses which disappeared again as the reflex was strengthened. During development of extinction, flashes again induced cortical responses. It is evident that, here, we are again dealing with an increase in cortical activity during internal inhibition.

Similarity of electric characteristics of the cortical region of the analyzer during positive and inhibitory reflexes was distinctly shown by Dolbakyan (1978). The evoked potential from the auditory cortex of dogs was markedly changed in shape when clicks became the signal of the motor defense reaction. It was shown in these experiments that the greatest rearrangement of evoked potential occurred in response to the first click irrespective of whether the delay reflex to a series of clicks or a short-latent reflex were elaborated. In both cases the late negative phase was strengthened and additional waves appeared in response to the first click. Shumilina (1966) observed the complication of the late waves of the evoked potential in specific nuclei of the visual system of the rabbit (external geniculate body, occipital area of cortex), both in response to the positive and inhibitory light CS. She explains these phenomena in terms of the reverberation of excitation in cortico-subcortical structures.

We have carried out electrophysiological testing of the functional condition of the auditory system in cats with the elaborated nictitating positive and inhibitory conditioned reflexes to sound stimuli (Gassanov, 1972). Electrodes were introduced into the region of the round window, cochlear nucleus, inferior colliculi, medial geniculate body and middle ectosylvian gyrus of the cerebral cortex. In all experiments, thresholds of the primary response to clicks were determined first. Threshold responses were determined to sound signals of increasing intensity, with a spacing of 2 dB, and stimulation was canceled when the threshold value was attained. The intensity of the conditioned stimuli was 50 dB higher than the threshold of hearing in humans. Sound signals 15 s long with a frequency of 10/s were confirmed by blowing air into an eye.

After the nictitating reflex to a series of clicks was elaborated, the thresholds of the primary response to the CS were decreased at all levels of the auditory system, except the responses of peripheral nerves. Conditioned reflexes were subject to extinction to a certain level: there were inhibitory reactions to a sequence of 10 presentations of a series of clicks. The thresholds of evoked responses were preserved at all levels of the auditory system, except the cerebral cortex. The threshold was further decreased in the cortical part of the auditory analyzer. When the process of extinction was lengthened, the thresholds of cortical responses were raised and the animal fell asleep. This suggests qualitatively different functional conditions of the cortical cells at different stages of extinction. In the case of extinction, the criterion of inhibitory reaction depth must be strictly observed since the identical effector index (inhibition of the elaborated reaction) might be accompanied by completely different neurodynamic processes.

In addition to extinctive inhibition, we studied the threshold of cortical primary responses to a differentiating stimulus (noise signals 1.4 ms long). The results were similar to the data on extinction: after differentiating inhibition was strengthened, the thresholds of cortical primary response were consistently decreased.

While considering the thresholds of primary responses as a method of direct functional testing of the excitability of the nerve cells, the following conclusion may be drawn. Elaboration of both positive and inhibitory conditioned reflexes is accompanied by a high level of excitability of all elements of the analyzer system, except its periphery. Internal inhibition is characterized by a higher level of excitability of the cortical region of this system.

Even in the first studies of neuronal activity, attention was drawn to the ratio between excitatory and inhibitory types of firing. It should be stressed that elaboration of all conditioned reflexes induces, as a rule, rearrangement of activity in a certain number of neurons, on average in half the recorded cells (Rabinovich, 1975). In the now classical work by Jasper et al. (1962) on monkeys with a motor defense reflex to light reinforced by electrical stimulation, 77% of motor cells increased their firing rate while 23% were inhibited in response to the CS. In response to a differentiating stimulus (light of another frequency) 61% of cells produced an excitatory reaction while 39% exhibited an inhibitory reaction. In the CS projection, inhibitory processes predominated both in response to positive and differentiating signals. No relationship was found between the evoked reactions of auditory neurons in monkeys or specific behavior in response to the positive and inhibitory sound stimuli (Goldstein et al., 1982). According to Rabinovich (1975) about half the neurons in the cortical projection zones of unconditioned and signal stimuli (sound plus electroshock in rabbits) responded to the CS by activation, while the other half responded by inhibition. Rabinovich correctly believes that the ratio between different forms of neuronal reactions could be due to the motivation basis of elaborated reflexes, the modal pattern of stimulus and the structure under study.

Rabinovich proposes the original hypothesis that inhibitory reactions of neurons in all brain structures present one of the forms of conditioned inhibition in response to the positive signal, since they are elaborated in response to combinations of the conditioned and unconditioned stimuli. The biological meaning of cellular inhibition might be seen in the realization of coordination and limitation of excitatory processes. Cell excitation in response to the inhibitory signal is viewed as a counteraction to the redundant inhibitory condition.

The data on the impulse activity of neurons in the association cortex are also of interest. Kubota et al. (1974) elaborated a motor food reflex to light in monkeys with the reinforcement delayed up to 9 s (light stimulation lasted 0.5−1.5 s). In the prefrontal cortex 63 out of 91 studied neurons were activated by light, and of these two-thirds continued to be excited during the period of delay through to the motor reaction. Similar results have been obtained in experiments with the prefrontal and lower parietal cortex of monkeys using a delayed reflex (Fuster and Alexander, 1971; Fuster, 1982). Watanabe (1986) concludes that, in the prefrontal cortex of monkeys, neurons are activated in both excitatory and inhibitory "behavioral mechanisms".

It is interesting that, in addition to observations suggesting a roughly equal ratio between the excited and inhibited neurons during positive and inhibitory reflexes, there is direct evidence of an increase in the excitability of the cortical neurons during internal inhibition. Analysis of ongoing and induced impulse activity of neurons in response to the CS suggests that inhibitory conditioned reflexes are characterized by a high excitability of the cells in projection zones of the cerebral cortex (Storozhuk, 1986; Vartanyan and Pirogov, 1986).

A long-term increase in the excitability of the sensorimotor neurons in cats was described after the nictitating reflex was elaborated and during its extinction (Brons and Woody, 1980).

It is evident that the ratio between the functional conditions of nerve cells in the cerebral cortex, where there is effector reaction, cannot be the same, even when neurons with inhibitory reaction predominate in the case of inhibitory learning. Neurophysiological analysis of brain activity has provided enough material to state that explaining integral behavior in terms of the condition of nervous system units is not adequate from the methodological point of view. The tendency to consider all conditions of nerve cells as the functional basis of a single reflex act is quite credible.

Quite distinct variations in the function of pyramidal neurons has been demonstrated in monkeys after learning (Evarts, 1973). Neurons changed their activity in time and space as part of the pattern of muscle activity and the magnitude and direction of force of the muscle of a working limb.

In the literature dealing with the electrophysiology of learning, internal inhibition is rarely treated as an independent problem. The main task is usually elucidation of the neurophysiological mechanisms underlying the formation of temporary connections, often with careful analytic discussion of the bioelectrical reactions obtained. This refers to a great extent to studies of cellular mechan-

isms of the conditioned reflex. In these studies, extinction and
elaboration of differentiation, in addition to pseudoconditioning,
became the classical methods of identifying the true conditioned
relationship and are not specially discussed. Nevertheless, the available
data concerning cellular reactions to inhibitory conditioned
stimulus allow the comparative analysis of neuronal activity during
positive and inhibitory conditioned reflexes.

It might be concluded that the excitatory nature of internal
inhibition can be proved at almost all levels of analysis, from goal-
directed behavior to neuronal activity. Moreover, some experimental data suggest that internal inhibition is based on the classical
principle of a combination of two stimuli: CS and unconditioned
signal. The latter differs from the usual reinforcement in that it does
not exist independently but arises only with respect to the previously
applied reinforcement, whereas internal inhibition is "reinforced",
as a rule, by cancellation of reinforcement. This makes the inhibitory conditioned reflexes biologically significant. They can be distinguished as a special class of conditioned reflexes which are excitatory
in their nature and which exhibit antagonistic behavioral reaction to
a certain previously elaborated positive reflex. According to this
principle, the inhibitory conditioned reflexes could be called negative and divided, according to their origin, into the same generic
classes as the positive reflexes.

It should be stressed that similarities between conditions of elaboration of internal inhibition and the positive reflexes suggest the
common neurophysiological bases of the two reflexes. And the term
"internal inhibition" acquires exclusively phenomenological importance. This latter opinion agrees with numerous statements by other
authors (Konorski, 1970, 1973; Kogan, 1979).

3. LOCALIZATION OF INTERNAL INHIBITION

Recently, researchers have more often paid attention to the search
for the localization of long-term memory. In an extensive review of
experimental studies on the neurophysiology of learning (Farley and
Alkon, 1985), the problem of localization of primary mehanisms of
the acquired reflexes in the nervous system was discussed in detail.
It was concluded that careful analysis of excitation circulation in

sensory, motor and interneuronal pathways activated during learning might be one of the approaches for solving this problem. According to Thompson *et al.* (1983), the search should be directed at revealing neuronal plasticity ("memory trace" in their terminology) in which learning would be encoded.

The problem of localization of internal inhibition is usually related to the search for the source of inhibition in the brain. Recognition of the excitatory nature of internal inhibition changes the direction of the search. The common principles of formation of positive and negative conditioned reflexes determine the common strategy of the search for the source of all reflex types. The structural basis of positive reflexes is investigated using methods revealing the degree of involvement of the studied nervous structure in conditioned activity. This approach can also be used for negative conditioned reflexes.

Recognition of the concept that cancellation of reinforcement induces an antagonistic reaction directs our research to the study of structures whose activity is reciprocal to the reinforced reflex structures. Konorski (1970) specially developed experimental techniques for the elaboration of two antagonistic reflexes with different architectonics and different excitatory relationships using reinforcement and unreinforcement. He writes in one of his last works (Konorski, 1973) about an inhibitory hunger center in the medial region of the prefrontal cortex. Extirpation of this region led to disturbances of inhibitory food conditioned reflexes.

The literature contains an experimental analysis of the structural differences between antagonistic positive and inhibitory reflexes. Two structures of the brain motivation system are separately involved in the realization of reinforced and unreinforced reflexes (Duglas, 1973). Extirpation of the amygdala resulted in animals having difficulty developing simple food and defense conditioned reflexes, but they were also able to elaborate extinctive inhibition. Various forms of inhibition were weakened in hippocampectomized animals. It was concluded that the hippocampus is able to "generate internal inhibition". However, after the combined destruction of the amygdala and hippocampus everything that had been elaborated before the operation was preserved. Two stimuli were distinguished as fully as in the normal animals, with the difference that in the operated animals the initial period of learning was markedly longer.

Some researchers raised doubts about the idea of structural differentiation of reinforced and unreinforced reflexes. Reskorla (1973) recognized the common mechanisms of formation of the positive and inhibitory reflexes, i.e. involvement of reinforcement in both cases, but, in contrast to Konorski, he believes that they are not antagonists. This belief is based on experimental data suggesting the mutual strengthening of reactions to positive and negative signals. Antagonistic reaction to an inhibitory signal could be explained, as Reskorla believes, by the activation of generalized reactions previously inherent in the conditioned signal. In this concept the degree of activation of the trace processes of reinforcement by conditioned stimuli plays the leading role. Both conditioned excitation and conditioned inhibition arise as a result of formation of relationships between traces of corresponding signals and the unconditioned stimulus. The positive reflex appears as a result of excitation of traces from the unconditioned stimulus while the inhibitory reflex increases the threshold of activation of the unconditioned reflex trace, this effect being possible only if accompanied by the positive reflex. Hence, antagonistic reflexes are a result of the functional condition of structures of the unconditioned reflex, which unites them, rather than of activation of different unconditioned reflexes.

In essence, Duglas (1973) recognizes the role of the same reinforcement in the elaboration of positive and inhibitory reflexes by centering attention on excitatory and inhibitory influences of some structures of the motivation system. However, Duglas draws a conclusion which, from our point of view, could be the key to understanding internal inhibition. On the basis of experiments with combined destruction of the hippocampus and amygdala he stresses that they facilitate learning and, more important, that they are no longer required after they fulfill their function.

Elaboration and preservation of conditioned reflexes and, in particular, internal inhibition, require special attention, since, according to experimental data, they are determined by different manifestations of nervous activity. Dynamic rearrangement of active states of the brain during elaboration and strengthening of the conditioned reflexes were found in numerous electrophysiological studies. Usually, the first combinations of stimuli not only excite vast areas of the cerebral cortex but also activate numerous specific, unspecific and motivation brain structures. But, as combinations of CS and unconditioned stimulus are repeated, the distributed form of activity is gradually weakened and the zone of activation limited.

These changes in bioelectrical reactions and their concentration during elaboration of the conditioned reflex were observed according to EEG, EP and impulse activity (Beck et al., 1958; John and

Killam, 1959; Livanov, 1972; Thompson *et al.*, 1972; Gabriel *et al.*, 1977).

Brain structures were also diffusely activated during elaboration of internal inhibition (Van Tai-an and Nzelina, 1963: Livanov, 1972). Excitation accompanying the initial stage of extinction is well known from the experiments with classical and instrumental reflexes which demonstrate the wave-like elaboration of inhibitory behavior.

Two forms of spatial organization of brain activity during learning distinctly suggest different structural bases of nervous activity during the initial period of conditioning and after its strengthening.

If positive and inhibitory conditioned reflexes are formed against the background of high activity of the cortical and subcortical structures, the preservation of the elaborated reaction appears to require the selective activity of cortical structures, namely the cortical projection area of the CS.

This conclusion is based on numerous facts, firstly on the results of experiments with extirpation of the cortex; this problem was investigated in detail in Pavlov's laboratories (Maiorov, 1954). The general conclusion was that, in animals with full or partial decortication, both positive and inhibitory reflexes were preserved and even new ones developed. It is important to note that the operated animals were, as a rule, markedly inferior compared with the control ones in terms of fine differentiation, accuracy of reactions, and rate of elaboration.

The concept concerning cortical localization of the mechanisms of learning is based not only on numerous studies of extirpation of various cortex regions, but, to a greater extent, on experiments with so-called transfer. Already by 1917 Frideman had shown that differentiating inhibition elaborated within the system of food conditioned reflexes is preserved in the system of acid conditioned reflexes as well. Reskorla and Lo Lordo (1965) elaborated the escape reaction in dogs, making them jump across a barrier in a box. In another situation, the classical conditioned reflex was then elaborated by combining one tone with an electrocutaneous stimulus, while another tone was presented without reinforcement. After many presentations of the reinforced and unreinforced signals the dog was again placed in the avoidance situation and sound CSs were applied against the background of stable instrumental reactions. The positive signals accelerated the avoidance reaction while the negative, differentiating signal markedly decreased the rate of reactions. Since in the experiments carried out by Reskorla and Lo Lordo, the classical conditioned reflexes had no effector manifestation, two motor conditioned reflexes were elaborated in experiments carried out by Weisman (1969). Rats were taught to rotate a wheel with electrical reinforcement and conditioned inhibition was then elaborated in a situation where the animals fled from punishment. Inhibition elaborated in the experiments with running was also effective with the wheel rotation reflex.

These results suggests that internal inhibition could be visualized both in experiments with classical reflexes and in those with instrumental reflexes. However, preservation of the inhibitory effect during different motor reactions is the most interesting (Weisman, 1969). In other words, internal inhibition elaborated during one form of behavior remained effective during another. Hence, cortical cells of the inhibitory stimulus perform their elaborated functions independently. However the fact that experiments with transfer of conditioned inhibition were carried out on reflexes, uniform in unconditioned stimulus, suggests that the quality of reinforcement and, consequently, of corresponding structures realizing the reflex may play an important part in transfer. This conclusion is not consistent with the results of experiments on food and rejected substances (Frideman, 1917). Later studies showed that inhibitory defense signals stimulate food instrumental reactions and vice versa (Reskorla and Solomon, 1967; Overmeier and Bull, 1970). This could be expected, if internal inhibition originated from cancellation of that unconditioned reflex which took part in the formation of the positive reaction. Specificity of reinforcement appears to leave a certain trace on positive and inhibitory conditioned reflexes elaborated on the basis of this reinforcement.

In experiments with transfer of uniform reflexes the role of higher parts of the analyzing (specific) system is quite distinct. The problem as to whether other subcortical brain structures are involved in mechanisms of preservation of the transferred reflex is solved by observations which have shown the presence of conditioned reflex in the case of full suppression of the unconditioned reflex (Gassanov, 1972).

Hence, behavioral experiments suggest that a two-step process of learning is crucial in the program of localization of internal inhibition. There is enough evidence pointing to localization of the end result of elaboration of internal inhibition in the cortical terminus of the analyzer. This is also confirmed by electrophysiological data. These results also suggest the similarity of neurodynamic processes during elaboration of positive and inhibitory conditioned reflexes. We carried out experiments with secondary extinction of motor defense reflexes in dogs (Gassanov, 1972). The technique of consecutive extinction of uniform reflexes is not new. Our aim was to combine the paths of realization of primary and secondary extinguished reflexes and to test the links of the inhibited path with the signal of secondary extinction. Sounds of one frequency differing in intensity by 10 dB (50 and 40 dB over the hearing threshold in man) were used as CSs. The forepaw was lifted and held during sound presentation, after which electrocutaneous stimulation of superthreshold strength was applied. After reflexes were strengthened, reaction to the strong sound was inhibited and the effect of weak sound without reinforcement was tested against the background of its inhibition. The reflex to the weak sound was preserved although it was shortened from the end. This effect was repeatedly observed but did not often restart the primary extinguished reflex. As expected, the second reflex was inhibited after a few trials.

A motor reaction to the weak sound with the same latent period as in usual experiments appeared against the background of the inhibited reflex to the strong sound. The curve of movement preserved its rectangular shape with a maximum of uplift despite the reduction of time during uplifting.

On the basis of these data the suggestion was put forward that the whole path of the inhibited reflex is free of any inhibitory process. Preservation of the response shape could be explained by preservation of the functional structure of the motor reactions. In other words, internal inhibition as an activity without physiological inhibition developed in the same links of reflex activity as the positive reflex.

A similar suggestion was put forward by Storozhuk (1986) when cellular reactions of different areas of the cerebral cortex and periaqueductal grey substance were studied in cats. In these experiments a motor reaction to a click with food reinforcement was elaborated. A series of clicks served as a differentiating stimulus. The number of neurons in the auditory cortex responding to both stimuli decreased and the neurons started to respond differentially. In the motor and somatosensory cortex, neurons responded to both stimuli but the reactions to inhibitory signals had a short latent period and were shortened. In periaqueductal grey substance only the activity was reduced in response to a differentiating stimulus. These results allowed the author to conclude that "extinction and differentiation take place in the same systems of excitatory synapses which are subject to basic rearrangement during learning (Storozhuk, 1986, p. 245).

On the basis of data obtained in studies of the neuronal activity of various brain structures of dogs using conditioned reflexes, Vartanyan and Pirogov (1986) concluded that nervous processes during the inhibitory reflex develop in the same elements as in the case of positive reflex and are similar in excitability. The localization of internal inhibition as one of the active functions of the higher parts of the nervous system coincides in essence with the concept of the structural foundations of the positive reflex. Electrophysiological and behavioral studies suggest that acquired activity, including inhibitory conditioned reflexes, is fixed in those sensory systems where the biologically important signal is addressed. The role of higher parts of the sensory system in mechanisms of learning and memory was confirmed by clinical practice and neuropsychological studies (Luriya, 1973; Glezer, 1985). Konorski (1972) stated that all kinds of memory are localized in the perception projection. According to Beritashvili (1968), the memory of a piece of food and its location is an exclusive function of the cerebral cortex in higher vertebrates. He distinguishes conditioned memory localized in cortical neuronal complexes which perceive combined stimuli and believes that images of both signal and reinforcement are present in conditioned memory.

All these ideas have been based on certain facts and stress as a whole the leading role of the cortical projection zones in preservation of acquired reactions. Comparing elaboration and preservation of negative, inhibitory conditioned reflexes with those of positive conditioned reflexes I have shown that internal inhibition also proceeds through stages of learning and fixation and that the latter stage is fixed as memory in the cortical cells which perceive a signal, i.e. a biologically significant stimulus.

4. MECHANISMS OF INTERNAL INHIBITION

The problem as to why cancellation of reinforcement transforms the previously positive signal into a negative one which ceases to induce an external effector reaction is the central and most complicated issue in analysis of neurophysiological bases of internal inhibition.

We do not use the term "inhibitory" intentionally since, in this formulation, the question already suggests an answer, linked to the search for inhibitory processes.

All the data considered raise the concrete question: why does activation in the cortical projection area, which appears due to omission of reinforcement, cease to induce the conditioned reflex? Indeed, if internal inhibition develops in the same structures which took part in realization of the conditioned reflex and if this process is excitatory, what mechanism deprives conditioned excitation of the ability to induce the motor reaction?

Electrophysiological studies of nictitating conditioned reflexes in cats allow us to conclude that elaboration of conditioned inhibition leads to limited excitation in the CS cortical projection (Gassanov, 1972). Let us consider the experimental facts which have led to this conclusion. First, the high level of excitability of cortical cells during extinct and differentiating inhibitory reflexes; the thresholds of evoked responses in the cortical projection of the negative signal were markedly inferior to those of the positive signal. In other words, a reaction to the negative signal appeared against the background of a new condition of cortical cells in terms of their activation. Second, if elaboration of the positive reflex was accompanied by a strengthening of late fluctuations of the evoked cortical response to the CS, these fluctuations were much weaker than in the same animals before elaboration of the conditioned reflex where there were negative signals. This inhibition of late waves took place against the background of high excitability, which meant that the observed changes could not be explained by the development of inhibitory processes in the cortical cells. Finally, in the motor cortex (in the projection of nictitating muscles) the development of negative reflex was accompanied by inhibition of all components of the evoked response to the inhibitory stimulus.

These results were later confirmed by some other studies. For example, Saunders (1971) has studied evoked response in the CS cortical projection zone in cats during elaboration of the instrumental defense reaction and concluded that early components of the evoked responses reflect the condition of "inputs" of the cortical cells while the late components reflect the condition of "outputs". The most sensitive to conditioning were the late responses whose changes were not related to the reactions of awakening since the evoked responses to background stimulus remained unchanged. Conditioned strengthening of the evoked response late waves could be due to the distribution of excitation along the established cortical pathways. However, the long-term realization of conditioned reflex ("overlearning" according to Saunders) led to a decrease in the amplitude of the evoked response. Amplitude rearrangement of the evoked potential appears to reflect different stages of brain activity considered in detail in the previous section. I have in mind the stages of generalized and selective activity when bioelectrical indices of learning are not only weakened but can disappear altogether.

The importance of later waves in realization of the conditioned reflex have also been demonstrated in experiments in which, after the conditioned reflex was elaborated, a short-term electrical stimulation was applied (John et al., 1968). If this stimulation was timed to the early components of the evoked potential, the conditioned reflex appeared in 59.5% of the CS applications. If the late components were affected by electrical stimulations, the conditioned reflexes appeared in only 20.9%.

The published data on inhibition of the late components during eleboration of the conditioned reflex and on their strengthening during extinctive inhibition do not, I believe, contradict this point of view. It is known that the spontaneous bioelectrical activity markedly affects the evoked response. Slow or desynchronized activity of the cortical cells can markedly enhance or weaken (until full abolition) the evoked potential. At the same time, the background rhythm is to a great extent determined by the strength and quality of the unconditioned stimulus, the conditions of potential recording, on freely moving or restrained animals. Hence, interference of two bioelectrical indices can result in both distinct and obliterated forms of the evoked potential. In addition, the most commonly used kind of internal inhibition, extinction, is usually applied without indications of the depth of its development. Meanwhile, long-term application of the stimulus can lead, as is well known, to a qualitatively different functional condition of the brain.

Studies with an additional control in the form of threshold estimations of the evoked potential or background stimulation with the other signals can apparently give more precise information about conditioned changes in the evoked potential components than comparisons of responses to the positive and extinct signals. This control liberated the evoked potential from the possible influence of rhythmic activity.

The mechanisms of functioning of the negative conditioned reflex without inhibition were shown in experimental studies in Livanov's laboratory (Livanov, 1981). These studies are characterized by analysis of the brain EEG on a new experimental–theoretical basis. Distant synchronization of the brain slow potentials was studied as a physiological index of excitation distribution in the cerebral cortex, as well as in subcortical structures. The conditioned reflex elaboration was accompanied by the strengthening of spatial synchronization of the cerebral cortex potentials. In the beginning the desynchronized activity was widely spread over the cortex and appeared during intervals between stimulations. As the reflex was strengthened, similarity between the cortex potentials was observed only between the projection zones of the combined stimuli and during the period of the CS action. The conditioned reflex extinction resulted in the complete disappearance of distant synchronization between the cortex projection zones.

A similar picture was observed during elaboration of differentiating inhibition. If light was combined with electrical stimulation of a hind paw in rabbits and sound was used as a differentiating stimulus, the CS increased coherence between the corresponding cortical areas, whereas the sound reduced it. During inhibitory reactions the processes occurring in different cortical centers were "uncoupled" and, as a result, the cortex conductance was affected (Livanov, 1972).

Dissociation of the activities of varius cortex areas during conditioned inhibition and disappearance of the conditioned reflex was also observed when studying evoked potentials (John et al., 1968). Rusinov (1969) believed that conditioned reflex has an obligatory stage of dominance, which is based on stationary excitation. According to the results of his experiments, the dominant condition in one cortical site changed functional relationships between distant sites, judging from the cross-correlation index. If the background biorhythms were not similar, it increased, and vice versa. At the initial stages of the conditioned reflex elaboration stationary excitation embraced vast cortical areas. After the conditioned reflex was strengthened, this excitation was better expressed at the CS cortical projection. This is further evidence of local manifestation of a nervous process during learning. This local process could be seen against the background of distant synchronization, if the experiments with polarization are taken into account.

In the light of these experimental data, the idea of local non-distributing excitation during negative conditioned reflexes appears as a physiologically real mechanism of negative learning. As to excitation, we usually have in mind the response to conditioned stimulus. It is obvious, meanwhile, that the conditioned stimulus should activate the neuronal organization fixed in the cerebral cortex in order to induce the elaborated reflex. The question arises as to why one neuronal organization in the projection zone of the conditioned stimulus induces an effector reaction, while another, which appeared in response to the inhibitory stimulus, does not induce this reaction. Why does excitation not spread beyond the cortical projection zone?

It might be assumed that inhibitory neuronal reactions, which arise in the cortex in response to both positive and negative signals, fulfill two functions. First, they organize impulse trains enhancing the most economic pathway to the effector cells. Second, they prevent, during internal inhibition, output to pathways which connect the afferent cells with the effector ones. However, the afferent properties of the cell, irrespective of the reaction type, suggest the spatial neuronal organization of activity of the analyzer cortical region, rather than differential involvement of the cells. It is possible that dynamic cortical organization determines the efficiency of the applied stimulus.

We studied interneuronal functional relationships in the auditory and motor cortex of cats during elaboration and extinction of the motor defense reflex to sound. The interneuronal relationships were evaluated according to cross-correlation analysis of impulse trains. The background multineuronal activity was treated with the aim of studying the trace processes underlying learning (Gassanov et al., 1985). Due to multiple recording via permanently implanted electrodes, the activity of both neighbor cells (under one electrode) and distant cells (under two electrodes) could be recorded in each cortical area. Before learning, relationships between neighbor neurons were observed much more frequently than between distant ones.

During elaboration of the conditioned reflex, the percentage of cells working in coordination in the auditory cortex markedly increased, in both neighbor and distant neurons. New connections developed between the distant neurons in the motor cortex. At the stage of strengthened reflexes all kinds of correlated activities of the cortical cells were preserved and even somewhat strengthened. The integrative activity of the cortical neurons was rearranged against the background of elaborated extinctive inhibition. The percentage of neuronal pairs acting in coordination at a distance markedly decreased. This was true of both auditory and motor cortex, but, in the auditory cortex, a strong link between neighbor neurons was preserved.

Studies of the correlated activities of cortical neurons convincingly suggest transformations of spatial integration of the neurons as a function of biological importance of the signals, spatial transformations being observed predominantly in the CS projection. A wide network of connections characteristic of the motor reflex was replaced during internal inhibition by detailed (in individual cortical microregions) distribution of interrelated neurons. It is possible that distant projection relationships determined the output of excitation from the auditory cortex into the motor cortex. Probably this function was fulfilled by the same relationships in the motor cortex. It is important that conditioned systemic activity in microclusters of neurons increased in the auditory cortex and suffered no changes in the motor cortex. Studies of systemic neuronal activity during learning were continued in our laboratory by Galashina and Bogdanov (Gassanov, 1987). They elaborated a motor food reaction to time in cats and analyzed correlations of neurons in the cortical projection of the working paw. The reflex elaboration was difficult with these animals, as seen from vegetative disturbances and the appearance of unusual reactions (long-term hygienic reflexes).

The results of these experiments have shown that, during elaborated reflex to time, when a motor reaction appeared during the last seconds of a 2 min interval, the number of correlated pairs of neighbor neurons reliably increased. But most demonstrative and highly significant were changes in the frequency of functional relationships between those neurons which were at a distance of 90 to 400 µm from each other (according to the distance between the recording electrodes). During the second minute the number of these relationships was three to seven times that during the first minute.

When the animal did not pause and often checked the food box with a paw, this tendency disappeared. Quantitative indices of correlated neurons in cortical microregions were leveled off, although at a high level, and interneuronal relationships at a distance were preserved at a low level. These data attracted attention to the fact that conditioned distribution of relationships was not induced by motor activity but rather determined by the presence of a reflex elaborated to time. Relationships between distant neurons were observed in an insignificant number of pairs during the initial period and were most intensive during the final period of preparation for the reflex to time.

As for mechanisms of formation and the functioning of inhibitory reflexes, the following opinion could be forwarded on the basis of experimental data: the first cancellation of reinforcement activates all structures involved in the formation of the positive conditioned reflex. These include specific systems of signal stimuli, activation and motivation formations and executive systems. A high level of activation enhances intensive interactions between different excited brain regions. This might be expressed in motor anxiety and strengthening of the elaborated reaction. Due to influences from the brain reinforcing structures induced by cancellation of usual reinforcement, the brain integrative activity is gradually rearranged, predominantly in the cortical projection of the negative reaction signal. As the latter is repeated without reinforcement, a new neuronal organization in the cortical projection is established. The new systemic activity is active, forms against the background of high excitability and, with its functional structure, limits excitation to the cortical microregions. The final formation of the conditioned-inhibitory cell ensemble proceeds against the background of marked weakening of the activity of structures involved in the initial stage of inhibitory conditioned reflex.

5. GNOSTIC NEURONAL NETWORK

We have considered three main problems of internal inhibition: nature, localization and mechanisms. All conclusions have been drawn from experimental studies and appear well founded within the framework of the analyzed published data.

The aim of this review is to provide evidence of the idea that internal inhibition is an excitatory activity of the brain, is localized in the cortical projection (for the signal) zone and is local. The two latter postulates coincide with the classical concepts of Pavlov, to which is added a qualitatively new understanding of the nature of brain activity, i.e. inhibition. This activity appears as a result of the organized network function of the cortical cells, which differs from the same function of the cortical cells during positive reflex by a marked reduction in relationships between distant projection neurons. Both types of conditioned behavior are characterized by a high level of correlated activity of neighbor neurons in cortical microregions. Using Konorski's terminology (1970), it can be proposed that the system of microgroups of neighbor cortical neurons fulfills a gnostic function while the system of distant neurons fulfills a communication or transient function. In our formulation, the concepts "gnostic" and "transient" are ascribed to correlated neurons, i.e. neuronal networks, rather than to individual neurons, as Konorski believes.

Discussion of the mechanisms of learning at the level of the neuronal network transfers elementary processes of nervous activity into a new category of characteristic values. Any condition of individual nerve cells might be implicated in the organization of the neuronal network, whose function consists in execution of a certain goal-directed behavior.

The idea that higher mental functions of the brain are based on integrative nervous activity both at the level of its structural formations and of its elementary units has a long history and is widely used in neurophysiological and psychological literature. Concepts of neuronal clusters could be found in Pavlov's studies (1951). He determined the structural dynamic complex of neurons and thought that they had a certain importance during elaboration of the conditioned reflex. Asratyan (1970) developed and formulated concepts concerning local conditioned reflex and local and conditioned state which are formed in cortical projection zones and expressed in the establishing of new, or the strengthening of existing, interneuronal functional relationships. Systemic activity (Anokhin, 1968) and image activity (Beritashvili, 1968) are based on neuronal clusters of various complexity. The concept of ensemble activity of neurons is a product of the pressing search for neurophysiological expression of higher brain functions, but needs physiological clarification in many aspects.

I would like to stress that the neuronal network is a selectively distributed system of relationships rather than a sum of excited and inhibited cells. It fulfills a cognitive function and is a material (cellular) basis of memory.

In neurophysiology, there is as yet only one (in availability and objectivity) method of studying correlated or net-like organization of neurons. This is the correlation method of analysis of statistical relationships between impulse trains. The most important results obtained using this method are, in my opinion, the following. First, the net-like function of neurons is not determined by individual properties of nerve cells. Such indices as reactivity, mean firing rate and distribution of intervals cannot be used to predict relationships between neurons (Gassanov, 1981, 1987). Second, systemic activity of neurons is directly related to learning. Of all quantitative indices of neuronal activity, their correlations might be considered the most stable.

This may be illustrated by an interesting study in which inter-neuronal correlations were analyzed in addition to the impulse activity of single cortical cells (Kogan, 1979). Experiments were carried out on wakeful cats in which the background and evoked firing rates were recorded in the visual and sensorimotor cortex at different stages of elaboration and fixation of a motor conditioned reflex. Conditioned changes in individual characteristics of neuronal activity, such as mean firing rate, coefficient of variation of inter-pause intervals, neuronal reaction type (phasic or tonic), distri-bution of excitatory, inhibitory and non-reacting neurons, were short-term and reversible. Changes in latent periods of impulse responses were preserved over a longer time. Changes in correla-tions between impulse trains of neighbor neurons in response to the CS were most stable. Kogan concludes that this structure of inter-neuronal relationships might be typical of the memory engram and the study of this latter should be directed at the description of the neuronal network.

Rabinovich (1975) stresses that plastic rearrangement of nerve cells preceding the appearance of conditioned reflex characterizes the formation of neuronal networks. Correlation analysis of net-like activity of the cortical cells has revealed a number of basic features in the formation and functioning of neuronal networks. First of all, the systemic properties of neurons are distinguished with reference to the anatomical position of the cells. They interact more intensely in the cluster position than at a distance. This property, quite understandable in the morphological aspect, proved to be more complicated in the functional aspect. Studies have shown that stimu-lation of cats by sounds and electricity, triggered by discharges of cortical auditory neurons, induces in the auditory cortex various distributions of relationships between neighbor cells. Moreover, stimulation with clicks but given in different sequences also changes the systemic neuronal activity (Gassanov, 1981). Various inter-neuronal relationships arise in the same auditory cortex when com-bining sound with electrocutaneous stimulation. These differences appear at all stages of the combined application of stimuli: during the reflex formation when there is as yet no stable motor reaction and at the stage of strengthened reflex (Gassanov, 1981; Gassanov *et al.*, 1985).

Internal inhibition is accompanied by the formation of a new quite different pattern of interneuronal relationships (Gassanov, 1981, 1987). I think that these results might be related to the characterization of the gnostic function of neuronal micronetworks. This function manifests itself in the background firing rate and determines corresponding behavioral reactions to the signal. There are reasons to believe that gnostic micronetworks are formed due to influences from many brain structures, which are regularly excited by combined (or single) stimuli. Simonov (1979) believes that the conditioned reflex is elaborated due to convergence of four afferent excitations: (1) by the CS, (2) condition of requirement (hunger, thirst, etc.), (3) receptors of the reinforcing agent and (4) emotional condition. I think that excitation from effector organs might be added to these sources. This understanding of the organization of the gnostic neuronal network in the signal cortical projection shows the function of this network to be fixation of both images of stimulation in a given situation and in the interdependence of these images.

Speaking about the function of the neuronal network we usually have in mind its activation by a stimulus. However, relationships between the induced response and the neuronal network should, apparently, be considered as physiologically independent but morphologically related processes. According to Livanov (1972), synchronization of potentials reflects pathways of excitation fixed in the brain, which arise in response to the CS. In our studies, interneuronal relationships were revealed in the background pulse activity in the absence of a stimulus or in intervals during reflex to time. Together with Livanov, I believe that gnostic neuronal networks in the form of interacting neurons are an expression of complicated memory traces at the neuronal level.

Transitory relationships which appear between gnostic neuronal micronetworks have already been mentioned. It is suggested that excitation induced by a biologically important stimulus is spread along them. According to EEG studies, both intracortical and interstructural functional relationships are formed and fixed in the brain (Livanov, 1972, 1981).

Hence, the memory neuronal network seems to consist of three

spatially separated levels, independent gnostic microregions, local clusters of microregions and spatially separated systems (Gassanov et al., 1985; Gassanov, 1987).
Repetition of the acquired reflex strengthens and, possibly, improves the system of functional relationships.
Doubtless, these suggestions require additional experimental evidence but there are enough facts to distinguish and substantiate local learning, which is timed to that cortical zone which realizes the analysis of the signal's physical and biological properties rather than to any cortical zone. If the brain has specifically tuned nerve cells to recognize the signal's physical properties, it mobilizes neuronal clusters of varying complexity to determine the signal's biological properties.
Speaking in images, the brain works with its cells and thinks with neuronal networks. According to the concept developed in this paper, any form of learning, as well as internal inhibition, has in its neurophysiological basis a gnostic neuronal network specific in function and local in structure. During internal inhibition the gnostic network acts as if in a pure form, whereas, during positive reflexes, it is complicated by transitory interneuronal relationships. In both cases, the gnostic network is the central link in conditioned activity and preserved memory traces.
The relationship between the induced excitation and the activation of memory traces appears to be based on complex neurophysiological mechanisms which require special studies. In any case, neuronal mechanisms of signal afferent excitation and learning cannot be considered as identical, although there is no doubt about the functional relationship between these mechanisms.

REFERENCES

Amsel, A. (1973) Patterns of formation of the conditioned relationship. In *Mekhanizmy Formirovaniya i Tormozheniya Uslovnykh Refleksov* (Mechanisms of formation and inhibition of conditioned reflexes), pp. 297–315. Moscow: Nauka (in Russian)
Anokhin, P.K. (1968) *Biologiya i Neirofiziologiya Uslovnogo Refleksa* (Biology and neurophysiology of conditioned reflex). Moscow: Meditsina (in Russian)
Asratyan, E.A. (1970) *Ocherki po Fiziologii Uslovnykh Refleksov* (Essays

on physiology of conditioned reflexes). Moscow: Nauka (in Russian)
Beck, E.C., Doty, R.W. and Kooi, K.A. (1958) Electrical reactions associated with conditioned flexion reflexes. *EEG Clin. Neurophysiol.*, **10**, 279−291
Belenkov, N.Yu. (1985) Importance of traces of previous nonconfirmation of the stimulus for formation of inhibitory conditioned reflexes in rats. *Zh. Vyssh. Nerv. Deyatel'nosti*, **35**, 110−116 (in Russian)
Beritashvili, I.S. (1968) *Pamyat' Pozvonochnykh Zhivotnykh, Ee Kharakteristika i Proiskhozhdenie* (Memory of vertebrate animals, its characteristics and origin). Tbilisi: Metsniereba (in Russian)
Brons, I.F. and Woody, C.D. (1980) Long-term changes in excitability of cortical neurons after Pavlovian conditioning and extinction. *J. Neurophysiol.*, **44**, 605−615
Dolbakyan, E.E. (1978) Averaged induced potentials of auditory cortex during short-delayed secondary defense instrumental conditioned reflex in dogs. *Zh. Vyssh. Nerv. Deyatel'nosti*, **28**, 490−492 (in Russian)
Duglas, R.D. (1973) Again to Pavlov. In *Mekhanizmy Formirovaniya i Tormozheniya Uslovnykh Refleksov* (Mechanisms of formation and inhibition of conditioned reflexes), pp. 371−397. Moscow: Nauka (in Russian)
Eccles, J. (1964) *Physiology of Synapses*. Berlin: Springer
Evarts, E.V. (1973) Interrelationship between the motor cortex activity, smooth movement and pose fixation. In *Mekhanizmy Formirovaniya i Tormozheniya Uslovnykh Refleksov* (Mechanisms of formation and inhibition of conditioned reflexes), pp. 141−162. Moscow: Nauka (in Russian)
Farley, B.J. and Alkon, D.L. (1985) Cellular mechanisms of learning, memory and information storage. *Ann. Rev. Psychol.*, **36**, 419−494
Frideman, S.S. (1917) *Dal'neishie Materialy k Fiziologii Differentsirovaniya Vneshnikh Razdrazhenii* (Further materials on physiology of differentiation of external stimuli). Saint-Petersburg: Diss. Doklad (in Russian)
Fuster, J.M. (1982) Cortical neuron activity in the temporal organization of behavior. In *Conditioning Representation of Involved Neuronal Functions. Advances in Behavioral Biology*, vol. 26, pp. 293−306
Fuster, J.M. and Alexander, G.E. (1971) Neuron activity related to short term memory. *Science*, **173**, 652−656
Gabriel, M., Miller, J.D. and Saltwick, S.E. (1977) Unit activity in cingulate cortex, and anteroventral thalamus of the rabbit during differential conditioning. *J. Comp. Physiol. Psychol.*, **91**, 423−440
Gassanov, U.G. (1972) *Vnutrennee Tormozhenie* (Internal inhibition). Moscow: Nauka (in Russian)
Gassanov, U.G. (1981) *Sistemnaya Deyatel'nost' Korkovykh Neironov pri Obuchenii* (Systemic activity of cortical neurons during learning). Moscow: Nauka (in Russian)
Gassanov, U.G. (1987) Gnostic micronetworks of cortical neurons. *Zh. Vyssh. Nerv. Deyatel'nosti*, **37**, 634−641 (in Russian)
Gassanov, U.G., Merzhanova, G.Ch. and Galashina, A.G. (1985) Interneuronal relations within and between cortical areas during conditioning in cats. *Behav. Brain Res.*, **15**, 137−146

Glezer, V.D. (1985) *Zrenie i Myshlenie* (Vision and thinking). Leningrad: Nauka (in Russian)

Goldstein, M.A., Jr, Benson, D.A. and Hienz, R.D. (1982) Studies of auditory cortex in behaviorally trained monkey. In *Conditioning Representation of Involved Neuronal Functions*. *Advances in Behavioral Biology*, vol. 26, pp. 307–318

Jasper, G., Ricci, G. and Down, P. (1962) Microelectrode analysis of discharges of the cortical cells during elaboration of conditioned defense reflexes in monkey. In *Elektroentsefalograficheskoe Issledovanie Vysshei Nervnoi Deyatel'nosti* (Electroencephalographic studies of higher nervous activity), pp. 129–151. Moscow: Izd-vo Akad. Nauk SSSR (in Russian)

John, E.R. and Killam, K.F. (1959) Electrophysiological correlates of avoidance conditioning in the cat. *J. Pharmacol. Exp. Therapeutics*, **125**, 252–270

John, E.R., Rushkin, D.S. and Villegas, J. (1968) Signal analysis and behavioral correlates of evoked potential configurations in cats. In *The Mind: Biological Approaches to Its Functions*, pp. 101–130. New York: Interscience Publ.

Kogan, A.B. (1979) Problems of neuronal organization of memory engram. In *Neirofiziologicheskie Osnovy Pamyati, VII Gagrskie Besedy* (Neurophysiological bases of memory, VII Gagry talks), pp. 60–70. Tbilisi: Metsniereba (in Russian)

Konorski, Yu. (1970) *Integrativnaya Deyatel'nost' Mozga* (Integrative brain activity). Moscow: Mir (in Russian)

Konorski, Yu. (1972) Problem of memory in a physiological aspect. In *Gagrskie Besedy* (Gagry talks), vol. 6, p. 56. Tbilisi: Metsniereba (in Russian)

Konorski, Yu. (1973) Some ideas concerning physiological mechanisms of internal inhibition of conditioned reflexes. In *Mekhanizmy Formirovaniya i Tormozheniya Uslovnykh Refleksov* (Mechanisms of formation and inhibition of conditioned reflexes), pp. 241–256. Moscow: Nauka (in Russian)

Kotlyar, B.I. (1963) Bioelectrical activity of some cortical and subcortical structure during differentiating and extinctive inhibition. In *Tezisy i Referaty Dokladov 20-go Soveshchaniya po Problemam Vysshei Nervnoi Deyatel'nosti* (Abstracts of the 20th meeting on problems of higher nervous activity), pp. 142–143. Moscow-Leningrad: Izd-vo Akad. Nauk SSSR (in Russian)

Kratin, Yu.G. (1967) *Elektricheskie Reaktsii Mozga na Tormoznye Signaly* (Electrical brain reactions to inhibitory signals). Leningrad: Nauka (in Russian)

Kubota, K., Ywamoto, T. and Suzuki, H. (1974) Visuokinetic activities of primate prefrontal neurons during delayed-response performance. *J. Neurophysiol.*, **37**, 1197–1212

Kupalov, P.S. (1955) General results of studies of inhibition in cerebral cortex. *Zh. Vyssh. Nerv. Deyatel'nosti*, **5**, 157–166 (in Russian)

Livanov, M.N. (1972) *Prostranstvennaya Organizatsiya Protsessov Golovnogo Mozga* (Spatial organization of brain processes). Moscow: Nauka (in Russian)

Livanov, M.N. (1981) On functional importance of some subcortical structures. *Uspekhi Fiziol. Nauk*, **12**, #3, 3−21 (in Russian)

Luriya, A.R. (1973) *Osnovy Neiropsikhologii* (Bases of neuropsychology). Moscow: MGU Press (in Russian)

Maiorov, F.P. (1954) *Istoriya Ucheniya ob Uslovnykh Refleksakh* (History of theory of conditioned reflexes). Moscow-Leningrad: Izd-vo Akad. Nauk SSSR (in Russian)

Overmeier, J.B. and Bull, J.A. (1970) Influences of appetitive Pavlovian conditioning upon advance behavior. In *Current Issues in Animal Learning*, pp. 117−141. Univ. Nebraska Press

Pavlov, I.P. (1951) *Polnoe Sobranie Sochinenii* (Full collected works). Moscow-Leningrad: Izd-vo AN SSSR (in Russian)

Pressman, Ya.M. and Varga, M.E. (1960) On functional interaction of nervous structures perceiving conditioned and unconditioned stimuli during extinction. In *Tezisy i Referaty Dokladov 19-go Soveshchaniya po Problemam Vysshei Nervnoi Deyatel'nosti* (Abstracts of the 19th meeting on problems of higher nervous activity), vol. 2, p. 71. Leningrad: Izd-vo Akad. Nauk SSSR (in Russian)

Rabinovich, M.Ya. (1975) *Zamykatel'naya Funktsiya Mozga* (Closing function of brain). Moscow: Meditsina (in Russian)

Reskorla, R.A. (1973) Model of conditioned reflex formation after I.P. Pavlov. In *Mekhanizmy Formirovaniya i Tormozheniya Uslovnykh Refleksov* (Mechanisms of formation and inhibition of conditioned reflexes), pp. 25−38. Moscow: Nauka (in Russian)

Reskorla, R.A. and Lo Lordo, W.M. (1965) Inhibition of avoidance behavior. *J. Comp. Physiol. Psychol.*, **59**, 406−420

Reskorla, R.A. and Solomon, R.L. (1967) Two-process learning theory: relationships between Pavlovian conditioning and instrumental learning. *Psychol. Rev.*, **74**, 151−182

Rusinov, V.S. (1969) *Dominanta* (Dominant). p. 231, Moscow: Meditsina (in Russian)

Saunders, J.C. (1971) Selective facilitation and inhibition of auditory and visual evoked responses during avoidance conditioning in cats. *J. Comp. Physiol. Psychol.*, **76**, 15−30

Serkov, F.N. (1977) *Elektrofiziologiya Vysshikh Otdelov Slukhovoi Sistemy* (Electrophysiology of higher parts of auditory system). Kiev: Naukova Dumka (in Russian)

Shul'gina, G.I. (1978) *Bioelektricheskaya Aktivnost' Golovnogo Mozga i Uslovnyi Refleks* (Bioelectrical activity of brain and conditioned reflex). Moscow: Nauka (in Russian)

Shul'gina, G.I. (1984) On the problem of neurotransmitter mechanisms of internal inhibition. In *Tezisy i Referaty Dokladov 17-go Soveshchaniya po Problemam Vysshei Nervnoi Deyatel'nosti* (Abstracts of 17th meeting on problems of higher nervous activity), pp. 86−89. Leningrad: Nauka (in Russian)

Shumilina, A.I. (1966) Experimental analysis of cortical−subcortical reverberation of excitations during extinction of conditioned defense reaction using the method of induced potentials. In *Voprosy Fiziologii i Patologii Nervnoi Sistemy* (Problems of physiology and pathology of the nervous system), pp. 5−24. Moscow: Medgiz (in Russian)

Simonov, P.V. (1979) Physiological analysis of involvement of emotions in organization of behavior. In *Neirofiziologiya Emotsii i Tsikla Bodrstvo-vanie-Son* (Neurophysiology of emotions and of the wakefulness—sleep cycle), pp. 7–21. Tbilisi: Metsniereba (in Russian)

Skipin, G.V. (1940) On the nature of inhibitory (extinctive) process developing in higher parts of the central nervous system in dogs. *Trudy Fiziol. Lab. Akad. I.P. Pavlova*, **18**, 459–466 (in Russian)

Sokolov, E.N. (1970) Neuronal mechanisms of negative learning. In *Neironnye Mekhanizmy Obucheniya* (Neuronal mechanisms of learning), pp. 63–75. Moscow: MGU Press (in Russian)

Storozhuk, V.M. (1986) *Neironnye Mekhanizmy Obucheniya* (Neuronal mechanisms of learning). Kiev: Naukova Dumka (in Russian)

Thompson, R.F., Patterson, M.M. and Teylor, T.J. (1972) The neurophysiology of learning. *Ann. Rev. Psychol.*, **23**, 73–104

Thompson, R.F., Berger, T.W., Madden, J. and Watanabe, M. (1983) Cellular processes of learning and memory in the mammalian CNS. *Ann. Rev. Neurosci.*, 347–491

Van Tai-an' and Nezlina, N.I. (1963) On changes in electrical activity of cerebral cortex during formation of conditioned reflex in monkeys. *Zh. Vyssh. Nerv. Deyatel'nosti*, **13**, 235–243 (in Russian)

Vartanyan, G.A. and Pirogov, A.A. (1986) Neuronal organization of conditioned inhibition. *Fiziol. Zh.*, **32**, 699–707 (in Russian)

Watanabe, M. (1986) Prefrontal unit activity during delayed conditional go no go discrimination in the monkey. II. Relation to go and no go responses. *Brain Res.*, **382**, 15–27

Weinberger, N.M., Velasco, M. and Lindsley, D.B. (1967) The reaction between cortical synchrony and behavioral inhibition. *EEG Clin. Neurophysiol.*, **23**, 297–305

Weisman, R.G. (1969) Some determinants of inhibitory stimulus control. *J. Exp. Analysis Behavior*, **12**, 443–450

INDEX

Amygdala 15, 16

Biorhythms 8
Brain motivation system 15, 16

Cancellation of reinforcement 5-7,
   20, 21, 26
Cats 11, 13, 19, 21, 24, 25, 28
Cerebral cortex *passim*
 — auditory 11, 19
 — interneuronal relationships
   24-29
 — somatosensory 19
CS cortical projection 12, 17, 21,
   23-26, 29

Decortication 17
Dogs 7, 8, 10, 17, 19

EEG 9, 23, 29
Evoked potentials 9-11, 22, 23

Habit transfer 17, 18
Habituation 10
Higher mental functions 27
Hippocampus 15, 16
History 3-6
Hyperpolarization hypothesis 9

Learning *passim*

Memory trace 15, 29, 30
Monkeys 10, 12, 13
Motor conditioned reflexes 12, 18,
   19, 24, 28

Nictitating reflex 11, 13, 21

Pavlov, I.P. 3, 4, 26
Periaqueductal grey 19

Rabbits 12, 23
Rats 7, 18

Unreinforcement 6, 7

Sov. Sci. Rev. F. Physiol. Gen. Biol., Vol. 5, 1991, pp. 37–71
Photocopying permitted by license only

THE DOMINANT AND THE CONDITIONED REFLEX

R.A. PAVLYGINA

*Institute of Higher Nervous Activity and Neurophysiology,
USSR Academy of Sciences, Moscow*

ABSTRACT

The concept of the dominant as a stable focus of increased excitability in the central nervous system, which predetermines the pattern of the nervous responses, was proposed by A.A. Ukhtomskii. The dominant is characterized by (1) increased excitability, (2) stable excitation, (3) inertia and (4) the ability to integrate stimuli coming from the other centers. The results are given for my studies into natural dominant states of hunger and thirst and the model of dominant focus produced in the motor cortex by polarization with a DC anode. The dominant condition was tested, in addition to sound stimuli, by stimuli inducing their own reactions, for example blinking. As the dominant is specialized, the interacting nervous centers form forward and backward temporary connections, and reciprocal inhibition does not spread being limited only to certain centers.

The conditioned reflex develops rapidly when motivation structures are in the dominant state. Discontinuation of this state using food reward, water or polarizing current serves as a reinforcement. The role of the dominant in the formation of classical and instrumental conditioned reflexes in the natural behavior of animals was analyzed. The method of learning resulting from discontinuation of the dominant state, developed by the author, is basically new. This form of learning belongs to a special class of endogenous conditioned reflexes.

KEYWORDS: Dominant, discontinuation (disruption) of the dominant, polarization dominant, reciprocal inhibition, forward and backward connections, temporary connection, conditioned reflex, goal-directed behavior, motivation behavior, endogenous conditioned reflex.

1. BASIC POSTULATES OF THE UKHTOMSKII DOMINANT THEORY

The dominant was discovered by Vvedenskii (1951a, p. 139). In 1881, in his studies of the influence of electrical stimulation of the n. vagus on respiratory movements, he discovered that, during an increase in the n. vagus tonus, short-term stimulation of the other nerves affected respiration as if the n. vagus itself was stimulated, when various sites of the motor cortex were stimulated. Stimulation of the cortical areas immediately thereafter induced an effect characteristic of the previously stimulated site. These facts were interpreted by Vvedenskii (1951b, p. 202) as "distortion" responses when, during any stimulation, the reaction induced is not the one typical of the given stimulus but another altogether.

In 1904 Ukhtomskii (1950, p. 31) was preparing a dog for one of Vvedenskii's experiments with stimulation of the motor cortex and was surprised by the following phenomenon: in response to cortex stimulation, which usually induced movement of the forelegs, defecation took place. The physiological significance of this phenomenon was not clear at that time and Ukhtomskii attributed it "to memory". But in 1909 he started experimental studies of this phenomenon and summarized their results in a thesis *On Dependence of Cortical Motor Effects on Side Central Influences* (Ukhtomskii, 1950, p. 31).

In 1923–1927 Ukhtomskii formulated the basic postulates of the dominant theory. The dominant is a general principle of activity of the nervous centers: it forms the basis of constant reorganization of relations between centers. The principle of dominant is given in general biological, philosophical and behavioral aspects.

Ukhtomskii defines the dominant in the following way: "The dominant focus of excitation which predetermines, to a great extent, the character of current responses of the centers at a given moment" (1950, p. 164), "...a stable focus of increased excitability of the centers, whatever the cause, so that excitations coming to the centers again enhance (reinforce) excitation in the focus, whereas inhibition is widespread in the rest of the central nervous system" (1950, p. 165), "...foci of increased excitability form partly under the influence of internal hormones, partly under the reflex influence in the nervous system which, in virtue of their increased excitability, will react more readily to diverse and remote stimuli applied to the organism" (1950, p. 186). And, further, "stable enough excitation

... accumulates excitations from the most remote sources but inhibits the ability of the other centers to react to impulses which are directly related to these latter" (1950, p. 190). And his last definition of the dominant is as follows: "The dominant (from Latin dominare, dominate) in physiology is a temporarily dominant reflex which transforms and directs for a given period of time (all other conditions being equal) the activity of the other reflexes and of the reflex system as a whole" (1950, p. 325).

Characteristic properties of the dominant include stable excitation, increased excitability, summation and inertia.

Stable excitation in the center is generally due to long-term or repeated activation of the reflex arch. If a short-term stimulus is applied, excitation in the center is not long-lasting and no dominant is formed. A focus of stable excitation in the central nervous system can be created not only by the reflex but also "under the influence of endogenous hormones".

The dominant focus has an increased excitability. Ukhtomskii stressed that, for the dominant to be formed, increased excitability of the center is required. In the case of excessive excitability (hysteriosis), the excitation focus has no summation properties. Stable excitation and increased excitability allow the focus to react more readily to diverse stimuli, i.e. it acquires summation properties, which are among the main dominant's properties. The dominant is also characterized by inertia, which was defined by Ukhtomskii as "fast renewal" or "short-term survival of dominant" (1950, p. 202). All these properties of the excitation focus determine the dominant formation.

Let us consider another aspect of Ukhtomskii's dominant theory. His ideas concerning organization and coordination of reflex acts are directly related to the determination of behavior. He indicates that the reflex system, which dominates among many possible responses, determines the goal-directed behavior of an animal: "Any time there is a symptocomplex of the dominant, there is a behavioral vector determined by it. And it is quite natural to call it 'organ of behavior'..." (1950, p. 300). Further, "the dominant is not only a normal working principle of the center, but plays an important part in *de novo* formation of reactions to the environment" (1950, p. 192).

The problem of motivation can be traced in Ukhtomskii's work in its physiological and psychological aspects. While considering the dominant as a source of goal-directed behavior, he always supported the reflex theory. Ukhtomskii dominant theory, which rejected the constant mechanism with "unambiguous effect", seemed to break the traditional reflex scheme according to which receptor stimulation inevitably induces the same motor or secretory response. Nevertheless, like Sechenov and Pavlov, Ukhtomskii believed that his ideas about the role of the dominant in behavior refine more precisely the classical reflex scheme with an account of transformation in the central apparatus of the integral reflex act, rather than reject the reflex theory. He concluded that the concept of the dominant explains "a lot about enigmatic variation of the reflex behavior of man and animals in virtually unchanging environment" (1950, p. 317).

Ukhtomskii's position during the discussion about instincts indicated that he could not accept predetermination of behavior by innate incentives only. At that time many researchers believed that instincts resulted from the regulation of actions by the corresponding goal. Ukhtomskii wrote: "Having seven instincts in hand we cannot explain concrete actions" (1950, p. 308). "It might perhaps be best to follow I.P. Pavlov who, in his lecture in America, simply did not raise the problem of instincts as separate from the reflexes, but was saying 'instinct or reflex'. Quite true, the concept of instinct includes nothing more than reflex" (1950, p. 30).

Hence, Ukhtomskii considered dominant as the principle of interaction between centers and concluded that the dominating reflex plays the role of a "stimulatory level" in animal behavior. In his concept of the dominant Ukhtomskii comes very close to the problem of behavior motivation. Recently, this aspect of the dominant theory has attracted the increasing attention of physiologists.

The dominant is usually considered a phenomenon related to an excitation focus in the central nervous system, which is reinforced by any stimulus. The conclusion is that the dominant is devoid of specificity. Meanwhile, in his paper *Dominant and the Integral Image* Ukhtomskii (1950) stressed certain stages of the dominant development: when applying stimuli of various modality the initial stage of reinforcement transforms to subsequent stages when the dominant is elicited by a "specialized" or "adequate" stimulus, i.e. acquires specificity.

Summation of stimuli coming from other parts of the central nervous system to the dominant focus causes a response to a stimulus which did not previously induce this reaction. Hence, similarity with temporary connection is demonstrated. This similarity allowed some authors (Rusinov, 1972; Pavlygina, 1973) to consider the dominant a simple form of temporary connection.

The aim of this review is to analyze some aspects of temporary connection based on the dominant at the stage of its specialization and to investigate the given temporary connection with special reference to realization of the dominant as a "behavioral vector". Since animal motivation behavior is based on the dominant, studies of these questions are related to a wider problem, namely the role of the dominant in goal-directed behavior.

2. METHODS OF INVESTIGATION

We studied various types of dominant: defense, swallowing, motor ("polarization"), thirst and hunger. In an attempt to assess as accurately as possible the functional condition of the center to where the dominant-testing stimulus was directed, we used dominant-testing stimuli which induced their own unconditioned reflex. These experiments were carried out on rabbits.

An animal was put on a table or in a special box that limited head movements. To produce a defense dominant, electrocutaneous stimulation in the form of a single pulse or a series of rhythmic pulses (1/5 s) was applied to the left forepaw via silver cup electrodes. The pulse length was 0.1 to 0.3 ms. Both subthreshold and threshold currents were used, which induced limb movement. The electromyogram was recorded using silver cup electrodes.

In experiments with "polarization" dominant an excitation focus was produced by applying a DC anode ($15\,\mu A/mm^2$) to the sensorimotor area of the right hemisphere in the left forepaw zone. For this purpose the bone was thinned down to the lamina vitrea and an active electrode (a $2 \times 2\,mm$ silver plate) was attached by Noracryl. The second electrode (a $20 \times 10\,mm$ silver plate) was attached to the animal's right ear.

Swallowing dominant was produced by pouring water into the oral cavity of an animal through a steel tube (3 mm in diameter) implanted into the skull nasal bones in front of the frontonasal suture. A narrow channel was cut with a dentist's bore in the middle turbinated bone, so as not to interfere with the ethmoid bone and nasal septum. The tube was inserted into the channel, with its lower end protruded by 1–2 mm into the oral cavity at the boundary between the soft and hard palate, thus ensuring the swallowing of all water and not interfering with subsequent normal feeding. During the experiments, a plastic tube was put on the outer end of the implanted steel tube. For the water supply we used a solenoid connected to an electric stimulator. The amount of water fed into the oral cavity was determined by the time of uplift of a magnetic bar constricting the plastic tube, which was connected to a water vessel fixed 0.5 m above the table. The water dosage required to induce a single swallowing (0.8–1 ml) was determined by the parameters of the current applied to the solenoid. For a dominant focus to be formed in the "swallowing center", a serial supply of water with intervals of 10 to 15 s was used. Swallowing movements were recorded via a silver wire electrode implanted into the neck skin at the level of the thyroid cartilage.

The blinking reflex was produced by either direct stimulation of the eyelid m. rectus via midget cup electrodes or by stimulation of the eye with a flow of air. For this purpose the same device with the solenoid was used. The pressure of the air flow was regulated by a manometer connected to the system of tubes. A special piece was put on the outer end of the tube implanted into the nasal bone which was connected with two other metal tubes bent in such a way that air flow was directed at the eye surface. One eye was stimulated and the tube in front of the other eye served as a control. The tubes were fixed on the animal's head, thus allowing eye stimulation with the same parameters, despite the head movements. The blinking reflex was recorded with silver wire electrodes implanted into the upper and lower eyelids of both eyes. For EEG recording, Nichrome electrodes (0.5 mm in diameter) were implanted into the skull bones of previously scalped rabbits, or steel screws with a tip diameter of 0.75 mm were used. Electrical activity of the zone of swallowing in the cortex was recorded via electrodes implanted over the antero-lateral area of prefrontal cortex according to the stereotaxic co-ordinates A-10, L-5.

Electrodes were implanted into symmetrical sites of the skull over each hemisphere to record electrical activity of the cortical "blinking center" (P-1, L-12 to 13), as well as in the cortical "forepaw" zone. Electrodes were also implanted into subcortical formations related to the swallowing function (P-6, L-5, Z-17 and P-5, L-2.5, Z-12). In order to record potentials of subcortical centers related to blinking, electrodes were implanted at coordinates P-4, L-3, Z-13.2 (medial part of thalamic VPM nucleus) and P-7, L-1, Z-17 (NIII). Electrodes were implanted into the subcortical structures through the left hemisphere using a stereotaxic device STM-3 and a chart produced by Sawyer et al. (1954).

The electrical activity of the subcortex was recorded via double Nichrome electrodes (100 μm) in factory-furnished isolation or steel electrodes (200 μm) insulated with varnish, except the electrolytically sharpened end. The electrodes were connected with the recording unit via a microconnector attached to the skull by Noracryl. Monopolar recording of electrical activity was done using an indifferent electrode fixed on the ear or nasal bones. Electrode implantation, as well as all other operations, were carried out under pentobarbital anesthesia (40−50 mg/kg intraperitoneally).

The electrical activity of the brain, swallowing and blinking reflexes and limb myogram were recorded on a strip chart recorder and, simultaneously, on an eight-channel tape recorder (Nihon Kohden, Japan) for subsequent computer processing. Analog signals were fed via an interface to an analog−digital converter connected to an ES-1020 computer. Sample volume and quantization frequency (0.46 Hz for every channel) were set by the software. The duration of analyzed realization was about 3.9 s. The analyzed realization introduced into the computer memory was divided into three equal parts, for which estimates of spectral density functions, mutual phase and coherent functions for all pairs of channels in a range of 1 to 30 Hz were calculated according to the algorithm of "Fourier rapid conversion" (Frolov and Sokolov, 1977; Pavlygina et al., 1983c).

The location of electrodes in the subcortex was determined by subsequent morphological control.

3. FORWARD AND BACKWARD CONNECTIONS IN THE CASE OF THE DOMINANT

It is essential that, at the stage of specialization, the dominant has both forward and backward connections between the center, where a dominant focus has been produced, and a center to which a dominant-testing stimulus is addressed. In rabbits, the defense dominant was produced by electrodermal superthreshold stimulation of the left forepaw (Figure 1A). Application of 20 to 30 stimuli with an interval of 3 to 12 s was sufficient to produce a defense dominant focus in the CNS, which started to respond to sound by limb contraction. There was also limb contraction after a small amount of water was poured into the oral cavity, which induces swallowing (Figure 1B). Defense dominant was tested two to three times by pouring water, i.e. the dominant was transferred from the stage of intensification of different modal stimuli to the stage of its testing by one stimulus. Subsequent sound stimulation induced not only limb contraction but also swallowing movements (Figure 1C). When the defense dominant produced by superthreshold stimulation of the forepaw was repeatedly tested by the blinking reflex, subsequent electrodermal stimulation induced, in addition to the paw movement, blinking of the corresponding eye (Pavlygina and Malikova, 1978, 1985).

Backward connection between the centers was found with the swallowing dominant as well. Serial water supply with intervals between series of two to three minutes produced a dominant focus in the swallowing center. The swallowing dominant formed, as a rule, after three to four such series. After water was no longer being poured into the oral cavity, the application of a tone, rustling sound, and stimulation of cornea and paw induced swallowing. In other words, the excitation focus produced in the "swallowing center" revealed summation. After repeated testing of the swallowing dominant by one of these stimuli, activation of the dominant focus from its own receptive field (pouring water into the oral cavity) induced not only swallowing, but also the reaction characteristic of the testing stimulus, i.e. backward connection between the centers was found. It is characteristic that, when the swallowing dominant was tested by stimulation of the cornea, its own unconditioned reflex (blinking) was not inhibited. At the stage of the swallowing dominant specialization, a single introduction of water into the oral cavity induced not only swallowing but also blinking of the corresponding eye (Pavlygina et al., 1983b). When the swallowing dominant was tested by electrodermal stimulation of the forepaw, backward connection between the centers was demonstrated as well. The swallowing dominant was produced by repeated pouring of water into the oral cavity (Figure 2A). When this dominant was tested by sounds (Figure 2B) or electrodermal stimulation of the paw (Figure 2D), swallowing movements appeared. The absence of limb muscle contraction (Figure 2D) suggests the presence of

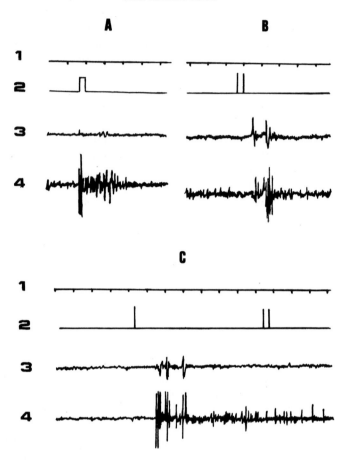

Figure 1 Forward and backward connections in the case of defense dominant. A, Formation of defense dominant by electrodermal limb stimulation; B, testing of defense dominant by pouring water into the oral cavity; C, subsequent testing of the dominant by a sound stimulus. 1, time control point, s; 2, stimulation; 3, swallowing; 4, limb electromyogram.

reciprocal inhibition characteristic of the dominant, since the stimulation of the forepaw was superthreshold (Figure 2C). The swallowing dominant was then tested from its own receptive field: pouring water into the oral cavity induced not only swallowing but also subsequent limb muscle contraction (Figure 2E).

These data are presented as a schematic diagram in Figure 3. When a dominant focus produced in the central nervous system

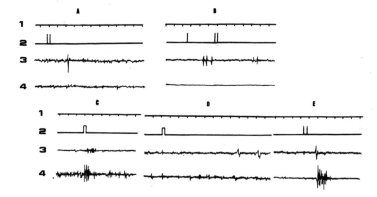

Figure 2 Forward and backward connections in the case of swallowing dominant. A, Formation of the swallowing dominant by pouring water into the oral cavity; B, testing of the swallowing dominant by sound; C, superthreshold electrodermal stimulation of the left forepaw before the formation of the swallowing dominant; D, similar stimulation of the paw against the background of swallowing dominant; E. pouring of a single portion of water. Order of recording as in Figure 1.

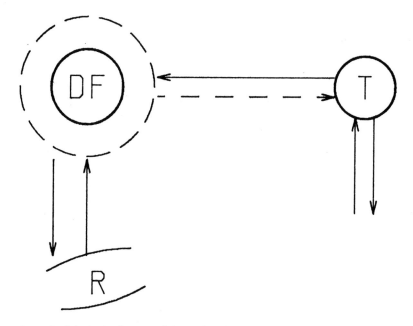

Figure 3 Schematic diagram of forward and backward connections between the centers in the case of dominant. For explanation see text.

(DF) is tested by another stimulus (T) and the latter induces the reaction characteristic of the dominant focus, we conclude that the relation between the centers is forward (solid arrow) by analogy with the conditioned reflex. If this dominant focus is repeatedly tested by the same stimulus, its subsequent activation from its own receptive field (R) induces in addition a reaction according to the backward connection (dashed line). A similar effect is induced by testing the dominant focus using any sensory stimulus.

When studying brain electrical activity, it was found that formation of forward and backward connections in the case of the dominant is reflected in the change of potentials of the corresponding centers. Ukhtomskii considered the dominant a constellation of centers linked into a single functional system. In order to estimate the contribution of different CNS levels to this constellation, electrical activity of the reflex cortical and subcortical links was studied during formation of the swallowing dominant (Pavlygina, 1985). Analysis has shown that the electrical activity of the dominant focus in the "swallowing center" has certain characteristics according to spectral-correlation parameters of the potentials. This change of potentials was especially distinct when dominant-testing stimuli were applied and at maximum before the swallowing was induced by summation of afferent inputs.

Spectral-correlation analysis of the electrical activity has shown that the dominant focus has a higher power in the range of delta-frequencies. Figure 4 presents spectrograms of the "swallowing center" potentials (solid line) in the cortex (I) and subcortical structures (II) 12 to 8 (1), 8 to 4 (2) and 4 to 0 s (3) before swallowing was induced by the tone. As the swallowing act is approached, the spectrum power in the range of delta-frequencies is increased in electrical activity of the "swallowing center" subcortical link, whereas there is as yet no marked change in the cortex. The power of delta-frequencies in the spectrogram of the cortical potentials increases only just before swallowing. Analysis suggests that subcortical links of the "swallowing center" are involved in summation earlier than the cortical ones. Spectrograms of the "blinking center" electrical activity (dashed line) did not manifest similar change of potentials.

If the cornea were repeatedly stimulated using air, to test the swallowing dominant, potentials of both "swallowing" and "blinking centers" would tend to increase in the power of delta-frequencies in response to the tone. Figure 4 shows that, 8 to 4 s before the summation reflex of swallowing, the power of potentials in the

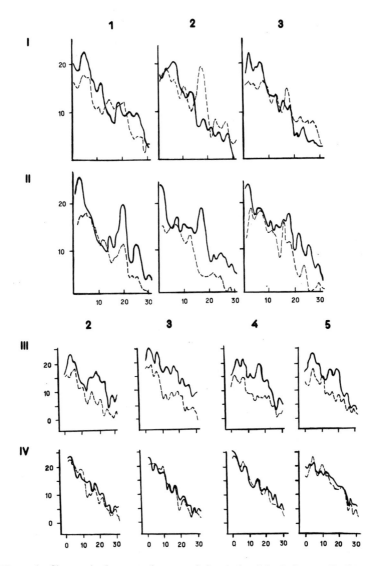

Figure 4 Changes in the spectral power of electrical activity before swallowing was induced by sound (tone). I and II, experimental testing of the swallowing dominant by sound only; III and IV, testing by cornea stimulation. I and III, spectrograms of cortical potentials; II and IV, spectrograms of subcortical potentials. Duration of analysis is 4 s. 1–3 represent 12 to 8, 8 to 4, 4 to 0 s before swallowing, respectively; 4 and 5 represent 4 to 8 and 8 to 12 s after swallowing. The solid line is the spectrogram of the "swallowing center" potentials; the dotted line indicates the "blinking center" potentials. Ordinate: spectral function; abscissa: frequencies.

delta-range (IV, 2) increases in the subcortical "swallowing" and "blinking centers". A similar change is then found in the electrical activity of corresponding cortical regions (III, 3). The increase in the power of delta-frequencies is more pronounced in subcortical links of the "swallowing center" and is preserved there 8 to 12 s after swallowing (IV, 5). A similar rearrangement of electrical activity in the "swallowing" and "blinking centers" in response to the sound stimulus after repeated testing of the swallowing dominant by eye stimulation reflects the formation of a temporary connection. Under the given experimental conditions, this appears as a backward connection between the dominating center and the center to which the dominant-testing stimulus was previously addressed.

Ukhtomskii defined the dominant as a system of functionally interrelated centers which determined the character of response. Linking of the centers into a single dynamic constellation reflects in unidirectional changes in the range of frequencies of the electrical activity. This constellation can be preserved for a long time and can be activated by other stimuli. These facts obtained both at the behavioral level and by analysis of brain electrical activity confirm Ukhtomskii's theoretical postulate. According to him, bidirectional relations between the centers appear at a certain stage of the dominant development: "A real ('adequate') connection is established between the dominant (internal condition) and a given receptive content (complex of stimuli), so that each counteragent (internal condition and external image) will exclusively induce or reinforce one another" (Ukhtomskii, 1950, p. 169).

Three subsequent phases in dominant development were distinguished by Ukhtomskii by psychological analysis of the behavior of certain characters in Tolstoi's novel *War and Peace*. Nevertheless, Ukhtomskii succeeded in putting forward physiologically substantiated postulates about changes in the dominant development, about the possibility of the dominant acquiring the specificity of bidirectional relations between the centers. In his notebooks we find further development of this idea formulated as a "law of backward connection between the centers": "If reflex A induces, by way of irradiation, another reflex B, vice versa reflex B will induce reflex A, by way of irradiation" (Ukhtomskii, 1966, p. 246).

The principle of bidirectional relation between the centers is well known in conditioned activity (Beritov, 1932). Pavlov (1951) suggested the movement of excitation in the reverse direction from the food center to the motor one and used these ideas to explain so-called voluntary movements. Since that time, the problem of conditioned backward connection has been actively developed by his students and followers (Kupalov, 1949; Asratyan, 1966, 1981; Struchkov, 1973).

Hence, two national neurophysiological schools (Pavlov's and Vvedenskii – Ukhtomskii's) discovered independently the law of backward connection, both during conditioned activity and in the case of the dominant which can be considered one of the forms of temporary connection. These phenomena are interrelated and the principle of forward and backward influences in all forms of temporary connections is universal (Asratyan, 1981).

4. RECIPROCAL INHIBITION IN THE CASE OF THE DOMINANT

In studies of various dominants using stimuli inducing their own unconditioned reflexes, the problem of reciprocal inhibition was touched upon. For example, in the case of defense dominant produced by stimulation of the rabbit paw, the "swallowing center" is not inhibited and the "blinking center" has an increased excitability. Against the background of the defense dominant, the blinking reflex amplitude is increased although parameters of the current stimulating the eyelid remain unchanged (Pavlygina and Malikova, 1978).

In the case of the swallowing dominant, the "blinking center" is not inhibited, while the limb movement center is in the condition of reciprocal inhibition. In our experiments, this was observed in the case of electrodermal stimulation of a forepaw, while Vasil'eva *et al.* (1966) obtained similar results after direct stimulation of the cortical zone of the forelimb projection. Superthreshold stimulation before the dominant formation became the subthreshold in the case of swallowing dominant. Studies of the blinking dominant (Drozdovska, 1987; Roshchina, 1988) produced by serial stimulation of a rabbit's eye with air have shown that the "swallowing center" is in the condition of reciprocal inhibition, while the forelimb movement center is not inhibited.

Hence, formation of a dominant focus in the CNS results in reciprocal inhibition of some centers and, apparently, in the inductive excitation of others.

What did Ukhtomskii think about reciprocal inhibition? In his works on the dominant, a certain evolution in ideas about reciprocal inhibition is evident. In 1923–1927 Ukhtomskii repeatedly stressed the diffuse character of reciprocal inhibition arising in the CNS during the formation of the dominant. Later, he became more cautious and emphasized that, here, reflexes functionally incompatible with the dominant are inhibited: "...any central response can become dominant and any response, which is hardly compatible with the prepared central activity, can be in the condition of reciprocal inhibition" (Ukhtomskii, 1936). Moreover, Ukhtomskii wrote: "the idea of reciprocal inhibition should not be distorted. This is not annihilation of all activity at specific sites but rather the processing of activity, its transformation more or less in agreement with the direction of dominating activity. Here quite local reflexes can be retained..." (Ukhtomskii, 1966). Of course, the "accursed" problem, as Pavlov called inhibition when studying the conditioned activity, is still open with respect to reciprocal inhibition in the case of dominant and requires further experimental studies.

5. GOAL-DIRECTED REACTIONS DEVELOPED ON THE BASIS OF THE DOMINANT DISCONTINUATION

Ukhtomskii repeatedly indicated that the dominant is "solved", or "completed", as a result of summation. Unfortunately, he did not specify how this process occurs, which, from our viewpoint, is especially important when the dominant realizes itself in goal-directed behavior.

All unconditioned reflexes are adaptive. A reflex abolishes, moves off, approaches, or prolongs the effect of the stimulus, which induced this reflex. When does this reflex become "dominant"? If no adaptive result is attained where the unconditioned reflex is realized (for example, defensive movement fails), the reflex is repeated many times, thus giving rise to a focus of stable excitation in the corresponding center. This is characterized by increased excitability and summation properties. The appearance of this excitation focus in the CNS leads to reorganization of intercentral relations, which can be represented as a call for help, an SOS, when the current activity is reinforced at the expense of other centers. Analysis of various dominant models suggests that completion of the dominant involves, first of all, removal of a source creating the dominant condition in the center. Long-term summation of excitations in the dominant focus without discontinuation of the dominant can lead to the formation of a stable pathological focus in the CNS and thus induce some diseases. The dominant, which is a "threshold of parabiosis" (Ukhtomskii, 1950, p. 273), transforms in these conditions into a pathological focus. For example, when studying the dominant formed as a result of the effect of direct current on certain hypothalamic regions, a model of arterial hypertension was produced in rabbits (Pavlygina, 1956).

The dominant plays an important role in the goal-directed behavior of animals (Pavlygina, 1979, 1980, 1982a; Batuev, 1982; Simonov, 1983). When motivational conditions arise, certain brain regions start to display the property of summation, i.e. the main property of the dominant (Hinde, 1970; Belenkov, 1979; Simonov, 1979; Sudakov, 1979). Due to summation of afferent inputs from the other CNS regions, excitation in the "motivation centers" attains the threshold level and triggers a chain of behavioral acts directed at satisfying an apparent need. Rapid realization of the dominating activity is of great importance for animal adaptive behavior (Pavlygina, 1982b). Motivation behavior of an animal (food, drinking, defense or sexual) directed at satisfying the need leads to the discontinuation of the dominant condition in the CNS. Analysis of experimental models of the dominant, as well as of their natural behavioral forms, suggests that animal behavior is directed at the discontinuation of the dominant state in the centers, rather than at its maintenance. This served as a theoretical prerequisite for experimental studies of the formation of temporary connection in conditions when a motivation dominant is formed by certain methods and then discontinued. After summation of dominant and afferent excitation achieved by any stimulus, we discontinued the stimulus producing the dominant condition in the CNS. Motivation thirst, hunger, "polarization" and defense dominants were studied (Pavlygina, 1982, a, b, 1985, 1987; Pavlygina et al., 1983a, 1987). In the latter case, when the zone of cortical presentation of a forepaw was stimulated by a DC anode, an excitation focus formed which had the properties of Rusinov's "polarization" dominant (Rusinov, 1957). Figure 5A shows that, after 10 minutes of DC application to the area of cortical presentation of the left forepaw, movements of the corresponding limb appeared in response to sound stimuli. After

A

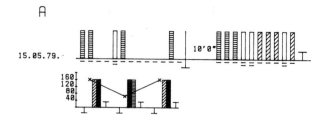

B

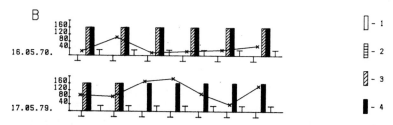

Figure 5 Graphic representation of the formation of goal-directed reactions on the basis of discontinuation of the "polarization" dominant. 1, Blinking reflex to an air flow directed at the eye; 2, general movement of an animal; 3, isolated movement of the left forepaw; 4, goal-directed blinking. ($-$), sound stimulation; ($=$), stimulation with an air flow; $\perp$, switching on of direct current; $\top$, switching off of direct current. Ordinate: time, s. Dots connected with a line designate the time of appearance of goal-directed blinking after switching on direct current. A, the first experiment with the dominant; B, manifestation of goal-directed reactions in subsequent experiments. Left: dates of experiments.

stable motor responses to the sound stimulus were established, eye stimulation induced the limb movement and the dominant state was discontinued by switching off direct current. On subsequent days (Figure 5B) the rabbit "voluntarily" regulated the dominant state: within a certain period of time (seconds, minutes) after switching on DC stimulation, distinct blinking appeared in the eye which was stimulated on the first experimental day, i.e. the rabbit produced the reaction directed at discontinuation of the dominant. "Voluntary" discontinuation of the dominant condition was regularly observed in all subsequent experiments. Analysis has shown that these voluntary responses are based on the backward connection activation. Eye stimulation tested the dominant focus at the stage of summation and, hence, the centers interacted according to the principle of forward temporary connection. Now, when the dominant condition appeared, the backward connection prevailed and activated the "blinking center".

Similar results were obtained when studying the motivation thirst dominant. Water deprivation (24 h) produced an excitation focus in the rabbit CNS. After repeated sound stimulation, swallowing and, sometimes, mastication appeared in response to sound (Figure 6A, I). At the optimal condition of the dominant focus, when summation became stable, eye stimulation with air induced unilateral blinking but, simultaneously, reinforced the dominant focus: the rabbit displayed swallowing (Figure 6A, II). Then the thirst dominant was discontinued, i.e. a small amount of water was poured into the oral cavity of the rabbit (Figure 6A, II). In subsequent experiments (with an interval of two to four days), where the thirst dominant was present, the rabbit discontinued the dominant condition. This was by blinking, in response to the sound, with the eye whose stimulation at the stage of summation induced swallowing and discontinuation of the dominant (Figure 6B, I). Since the rabbit produced the

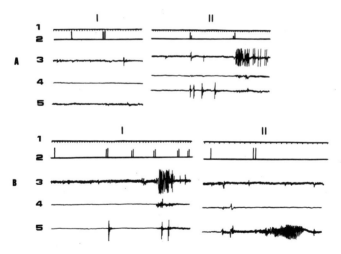

Figure 6 Formation of a goal-directed reaction on the basis of the thirst dominant. A: I, reactions to sound at the stage of summation; II, reactions at the stage of completion of the dominant. B: I and II, reactions of an animal directed at discontinuation of the dominant condition. 1, time, s; 2, control point of stimuli: switching on and off of sound (A, I; B, I; B, II), eye stimulation with an air flow and pouring of one portion of water (A, II); 3, recording of swallowing and mastication; 4, blinking reflex of the right eye; 5, blinking of the left eye.

reaction directed at discontinuation of the thirst dominant, it re-
ceived water. As seen in Figure 6B, I, swallowing and mastication
sometimes activated backward connection between the centers, as
expressed in blinking.

After a few experiments, this goal-directed blinking could become
tonic: the rabbit not only nictitated but closed its eye for a few
seconds (Figure 6B, II). Recording of myograms of the opposite
eyelids suggests that voluntary blinking is realized by one eye. If, at
the stage of summation, the right eye was stimulated, in subsequent
experiments the rabbit produced the voluntary reaction with the
right eye.

Let us consider changes in behavioral reactions in several experiments (Figure 7). In the beginning of the first experiment, a de-
prived animal swallowed relatively rarely in response to sound.
When such swallowing became stable, the cornea was stimulated
using air flow, thus reinforcing the excitation focus produced in
the CNS by water deprivation. The rabbit swallowed. Since the

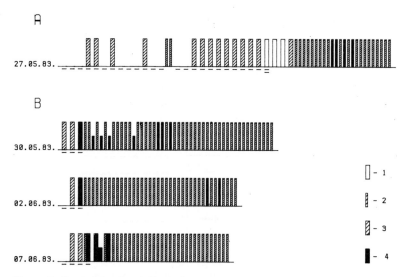

Figure 7 Formation of goal-directed reactions in the case of discontinuation of the
motivation thirst dominant. 1, Blinking of the left eye induced by stimulation of its
cornea; 2, swallowing after the pouring of water into the oral cavity; 3, swallowing
and mastication in response to sound; 4, goal-directed blinking. (−), sound stimulation; (=), stimulation with an air flow. A, the first experiment; B, subsequent
experiments. Left: dates of experiments.

dominant was discontinued after the summation reflex to the cornea stimulation, a few portions of water were poured into the oral cavity of the animal. Swallowing sometimes induced unilateral blinking with the eye which was previously stimulated. Goal-directed discontinuation of the thirst dominant by the animal was observed in all subsequent experiments. This reaction is directed at discontinuation of the dominant, as suggested by the data obtained with conditioned animals without water deprivation. The blinking reaction of one eye was, in these cases, absent. This implies that the voluntary reaction is based on activation of the backward connection between the centers: in the first experiment, stimulation of the "blinking center" reinforced the dominant focus and, in subsequent experiments, formation of the dominant focus activated the "blinking center".

The stage of dominant completion is the basis of rapid formation of a fast temporary connection. The goal-directed behavioral response based on the backward connection between the centers is elaborated usually after one (after two in one rabbit out of nine) completion of the dominant. Temporary connection between the centers was regularly observed in all subsequent experiments (Figure 7B), as well as in the experiments with "polarization" dominant.

I propose that, when there is water deprivation in conditions where the "drinking center" does not possess summation properties, a sequence of the same stimuli will result in rapid formation of an instrumental temporary connection in the rabbit. Experiments were carried out according to the following scheme. The animal was placed in a chamber where swallowing movements, mastication, and blinking reflex were recorded for 10 min without any special stimuli being applied. Sound stimulus was then applied and within 1 min (mean interval between dominant-testing stimuli) the eye was stimulated using air which induced blinking unaccompanied by swallowing, since summation was absent. The rabbit was then given water, i.e. the same experimental conditions were observed as with the dominant, although the motivation "thirst center" was not in a condition to sum up incoming stimuli. More than two dozen experiments were carried out on three rabbits but no temporary connection was elaborated. Individual blinking responses were observed in one rabbit after 12−14 experiments, but they were unstable. Comparison of these data suggests that the dominant condition in the center is the most adequate for formation of a fast temporary connection.

Similar results were obtained with the motivation hunger dominant. An excitation focus in the "hunger center" was produced by food deprivation (24 h). Sound stimuli induced mastication or, more rarely, swallowing (Figure 8A), i.e. the excitation focus produced in the "hunger center" was reinforced by excitation coming from the auditory analyzer. When summation reactions to sound became stable, air directed at the eye was used as a dominant-testing stimulus. If this stimulus induced a summation reaction in the form of mastication, the animal was given food and, hence, the motivation hunger dominant was discontinued (Figure 8A).

A single discontinuation of the motivation hunger dominant in all eight rabbits led to the formation of a stable temporary connection between the centers. In the next experiments, summation food reactions appeared after a few sound stimulations. The rabbit realized a blinking reaction with a corresponding eye and was rewarded with food. In all subsequent experiments discontinuation of the motivation hunger dominant was stable. Experiments were carried out twice a week, as when studying the thirst dominant.

A goal-directed reaction is formed also when excitation in the motivation center is not fully removed. Experiments were carried out with the motivation hunger dominant when, after summation in response to eye stimulation in the first experiment, the animal was given only a small amount of food. In this experiment, after the

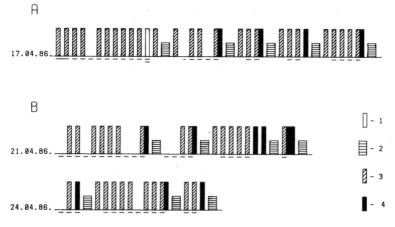

Figure 8 Formation of goal-directed reactions in the case of discontinuation of the motivation hunger dominant. 1, Blinking reflex of the right eye after its stimulation with an air flow; 2, feeding; 3, swallowing and mastication in response to sound; 4, goal-directed blinking. (−), sound stimulation; (=), stimulation with an air flow. A, The first experiment; B, subsequent experiments.

food box was closed, sound stimuli were again applied. The dominant was rapidly recovered and mastication appeared in response to sound. The rabbit then blinked, the food box was opened and a small amount of food was given to the rabbit. Hence, the temporary connection proved to be stable during repeated testing in the same experiment. The acquired behavior was preserved during subsequent experiments as well (Figure 8B). In some cases, blinking directed at discontinuation of the dominant became tonic: the rabbit closed one eye for a few seconds.

Let us visualize the formation of a temporary connection based on the dominant. Figure 9A shows the dominant at the stage of summation. The motivation center (of hunger, thirst) becomes dominant (DF) due to sound stimulation (upward arrows). The testing of the dominant with eye stimulation (T) also induces summation (downward arrow from the circle). But after this summation effect, either the dominant is discontinued (arrow with a rectangle) or the level of excitation in the motivation center is reduced (Figure 9B).

Next time, as soon as the excitation focus attained the summation state, the rabbit actively discontinued it by realizing the reaction

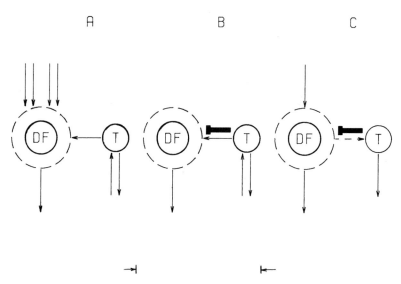

Figure 9 Schematic diagram of the formation of a temporary connection between the centers on the basis of discontinuation of the dominant. A, Dominant at the stage of summation; B, discontinuation of the dominant; C, expression of goal-directed behavior. DF, Dominant focus; T, center of the blinking reflex.

which had led to the dominant discontinuation (Figure 9C). Goal-directed reaction is realized at the expense of backward connection between the centers. Any motor reaction could be taken in place of eye stimulation. These facts confirm Pavlov's concept (1951) that the goal-directed reaction is based on activation of the temporary backward connection. This mechanism was proposed by Pavlov when explaining the instrumental conditioned reflex.

The goal-directed reaction can also manifest itself without sound stimulation, as suggested by the results obtained when studying the "polarization" dominant. During the DC stimulation the level of excitation in the limb center increases, the focus attains the dominant state and induces blinking. This model most closely simulates learning in nature, when excitation in the motivation center is increased, as a rule, by endogenous factors (humoral, hormonal), rather than by distant stimuli.

In nature the dominant state in the motivation hunger center is produced by changes in blood biochemical parameters and a search reaction appears. If the animal has not been conditioned by the parents and has no experience of its own, the search reaction is not goal-directed and food is searched for at random. Food appearance, smell, etc. reinforce the dominant state in the motivation center. This summation is replaced by discontinuation of the dominant state, i.e. eating. When the animal is next hungry, the complex of centers, which previously gave information about food, is excited by backward connection activation. Goal-directed behavior is aimed at a search for a certain object.

Hence, these experimental data concerning discontinuation of the motivation dominant suggest that if the motivation center is in the dominant state, the temporary connection, based on the given motivation, forms practically after one combination. In the presence of the motivation dominant (hunger, thirst), one combination of cornea stimulation with the other stimuli (food, water) is enough for a fast temporary connection to be formed. Next time, as soon as the motivation center attains the dominant state, the animal satisfies its need by realizing goal-directed reactions.

Adaptive reaction based on the dominant discontinuation is elaborated according to Pavlov's temporary connection principle. Appearance of a temporary connection between the centers follows from the fact that excitation of one center induces a reaction characteristic of another. The question arises as to why, when conditioned reflexes are formed, they appear after application of several combinations, rather than one.

6. THE ROLE OF THE DOMINANT IN THE FORMATION OF CONDITIONED REFLEXES

Pavlov and Ukhtomskii believed that the dominant plays a certain role in the formation of the conditioned reflex. For example, when explaining the mechanism of the conditioned reflex generalization, Pavlov wrote: "If we stimulate the center regularly, during the intervals this center is in latent excitation which does not manifest itself, but if you add another center of stimulation, irradiation from this latter reveals the excited state of the regularly stimulated center, whose effect becomes pronounced. This is what Ukhtomskii called the 'dominant' " (Pavlov, 1949, p. 21). In turn, Ukhtomskii proposed that "the 'dominant' is a key to explaining the mechanism of 'temporary connections' discovered by Pavlov in the activity of higher cortical reflexes" (Ukhtomskii, 1950, p. 193). The role of dominant phenomena in elaborating conditioned reflexes was repeatedly stressed later (Shnirman, 1926; Asratyan, 1941; Livanov and Polyakov, 1945; Skipin, 1951; Rusinov, 1957; Pavlygina, 1973).

Let us consider the elaboration of a conditioned food reflex (Figure 10). Pavlov believed that the main prerequisite of "the conditioned reflex formation is, in general, single or repeated coincidence in timing of the indifferent stimulus with the unconditioned one" (Pavlov, 1951, p. 451). At the same time, the conditioned food reflex is not elaborated in a satiated animal, i.e. hunger is mandatory for the conditioned food reflex to be elaborated. Hence, during elaboration of the reflex (Figure 10A) the combination of sound (S) with food (F) is preceded by the hunger condition (H). If, from the beginning of the reflex elaboration, food is given in all combinations, the motivation hunger state is reduced. It seemed as if, in these conditions, no motivation dominant could form but the stage of conditioned reflex generalization was present. Analysis has shown that the "hunger center" attains the dominant state (dashed circle in Figure 10B) under the influence of situational (SS) and conditioning (S) stimuli. If a dog is fed several times under experimental conditions, a powerful situational reflex forms. Salivation in response to any stimulus at the stage of generalization is an index of dominant focus in the motivation hunger center (downward arrow). There are data (e.g. Konorski, 1967) suggesting that, in the case of hunger, salivation is reinforced. Taste and smell sensitivity also increases along with all motor activity. If the dominant focus is tested by any stimulus, for example sound, the summation reaction will be realized

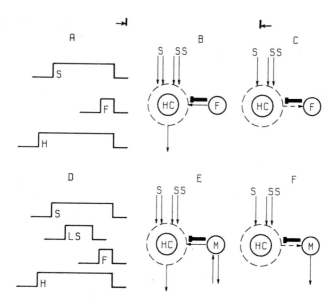

Figure 10 Schematic diagram of elaboration of conditioned food reflexes of (A-C) type I and (D-F) type II. H, State of hunger; HC, "hunger center"; S, sound stimulation; SS, situational signals; F, feeding; M, instrumental movements; LS, limb stimulation. For explanation see text.

only in response to this stimulus. At the stage of conditioned reflex generalization, when one stimulus is used as a conditioning signal, all conditions are created to transfer the dominant focus produced in the motivation center to the stage of specialization.

When the food box is open, afferent excitation induced by food appearance and smell undoubtedly reinforce the dominant state in the hunger center at the stage of conditioned reflex generalization as well, but this summation is replaced by the discontinuation of the dominant, since food is then given (Figure 10B, arrow with a rectangle). Eating is accompanied by salivation. There appears to be no fundamental difference if the dominant is gradually extinguished, as during elaboration of the conditioned reflex after a few combinations with food given in small portions, or at once, as when studying the discontinuation of the dominant.

Discontinuation of the motivation dominant leads to fixation of a temporary connection. The temporary connection becomes fast and operates in subsequent experiments, according to the backward connection (Figure 10C, dotted arrow) between the centers. Long-term summation of excitations in the dominant focus without its discontinuation (no food is given) leads to the focus inhibition, as expressed in the absence of "conditioned" response. When elaborating a conditioned food reflex with the constant duration of stimuli and intervals between them in combinations, conditioned reflex to time undoubtedly makes an important contribution to the conditioned response.

Hence, conditioned food reflexes are not elaborated after the first combination, since regular application of food in all combinations inhibits elaboration of a temporary connection, namely prevents the formation of the motivation dominant in the "hunger center", a prerequisite of rapid formation of a new reflex.

Elaboration of a conditioned defense reflex follows the same patterns. Motivation defense condition is produced by application of an unconditioned aversive stimulus to one of the animal's limbs. At the stage of generalization, formation of the motivation defense dominant is promoted by situational and conditioning stimuli. Since the defense motivation dominant was produced by limb stimulation, the stimuli applied induce its contraction. In the case of limb contraction in response to a conditioning stimulus at the stage of generalization, proprioceptive excitation reinforces the dominant condition of the motivation center but after this summation the current is not applied to the paw, i.e. the defense dominant is extinguished. Next time, when the motivation defense condition is recovered in response to sound, the paw center is activated by backward connection, the limb moves and the animal is not stimulated by current. There is no 100% realization of the conditioned response, since the absence of electrodermal stimulation leads to a decreased defense state. Backward connection is not activated, the limb conditioned movement is absent and, as a result, the animal receives electrical stimulation. This leads to the defense state recovery.

On the basis of this concept, I conclude that the rate of elaboration of conditioned reflexes is determined by the rate of formation of a dominant focus in the motivation center. Our experimental data suggest that discontinuation of the dominant state in the "motivation center" in the first experiment leads to rapid formation of a temporary connection.

How do the conditioned food and defense reflexes differ and what do they have in common? The motivation state, which is a prerequisite of the conditioned reflex elaboration, is produced by hunger during elaboration of the food reflex and by application of the second, in sequence, aversive stimulus during elaboration of the defense reflex. Elaboration of both reflexes proceeds through the stage of generalization which has the dominant properties. In both cases, irrespective of how the dominant condition in the motivation center forms, its discontinuation is a prerequisite of the conditioned reflex elaboration. The dominant condition in the motivation center is discontinued by feeding during elaboration of the conditioned food reflex or by removal of the aversive stimulus during elaboration of the conditioned defense reflex.

"Closing" is based on the fact that, after any activity (jumping to avoid something, salivation during conditioned food reflex, paw lifting during conditioned defense reflex) revealing the dominant state in the motivation center, this state is removed. Formation of a temporary connection takes place in fact in the first experiment with discontinuation of the dominant. In all subsequent experiments it is effective.

It should be noted that when discussing the conditioned defense reflex formation, we have in mind that an animal is able to avoid the painful stimulus, i.e. the conditioned reaction is truly adaptive. Another form of elaboration of the conditioned reflex is also used when the possibility of the dominant discontinuation is not envisaged: the animal receives electrical stimulation irrespective of its responses. Elaboration of the conditioned reaction in these experimental conditions appears to be related to the mechanism of waiting, so-called conditioned reflex to time.

If conditioned reflexes are considered from the point of view of the dominant, there is no essential difference between conditioned reflexes of types I and II. The motivation dominant appears to be produced during elaboration of conditioned instrumental reflexes as well. Figure 10D shows schematically the elaboration of the instrumental food reflex including the hunger state. An animal lifts a paw in response to sound (S) and then receives food (F). Conditioned reflex is not elaborated if these three stimuli are always combined. Combinations of sound and paw stimulation are necessary without food, thus increasing, as I believe, the hunger state which is removed by food. Situational stimuli also reinforce the hunger dominant (Figure 10E). After summation with afferent excitation induced by the limb movement, the dominant condition of the "hunger center" is gradually removed by food (arrow with a rectangle). If it is not fully removed, since a small portion of food is given, conditioned response in the form of paw lifting is induced by all subsequent sound stimulations. Here again, activation of backward connections between the centers is observed (Figure 10F). The results of experiments when both limb movement and salivation were recorded (Konorski, 1967, p. 368) suggest that the motivation dominant of hunger forms first and conditioned reaction appears thereafter. According to these results, salivation occurred on the fifth day after 28 combinations, in response to triple passive paw bending, suggesting the presence of the motivation hunger dominant. On the same day, the animal started to realize active paw movements.

Analysis of the conditioned reflex elaboration according to the method described allows us to give the following definition of the term "reinforcement": reinforcement is discontinuation of the dominant condition in the motivation center. During elaboration of conditioned food reflexes it is attained by feeding, during elaboration of conditioned defense reflexes by removal of the aversive stimulus. The conditioning stimulus reinforces excitation of the motivation center, thus inducing, in turn, according to the mechanism of backward connection between the centers, a conditioned response directed at discontinuation of the dominant condition in the motivation center. All types of conditioned reflexes are based on the presence of backward connection between the centers.

7. CONCLUSION

Analysis of various types of conditioned reflexes led to the conclusion that the formation of any conditioned reflex is based on discontinuation of the motivation dominant (Pavlygina, 1987). Since this is not usually taken into account during reflex elaboration, the dominant state is not controlled by an experimenter and, therefore, conditioned reflexes are elaborated only after a large number of combinations of associated stimuli. In nature such long-term learning could lead to the animal's death. If the dominant state of the motivation center is crucial for the formation of a temporary connection, the method of elaborating conditioned reflexes should be altered. The methods we use differ from those applied when elaborating conditioned reflexes of types I and II. The universally adopted combination in time of the conditioning stimulus with the unconditioned one is not used and much importance is attached to the influence on the endogenous condition of the motivation center and formation of a dominant focus in it. Discontinuation of the motivation dominant leads to rapid formation of a fast temporary connection. Elaboration of goal-directed reaction is completed in the first experiment and does not require repeated combinations of stimuli. These reflexes differ from operant conditioned reflexes (Skinner, 1953) although they are elaborated without applying combinations of stimuli. But formation of operant reflexes takes place without using methods which allow testing of excitation condition in the motivation center. The animal receives food after a random operant movement.

These reflexes, elaborated on the basis of the dominant, were called endogenous conditioned reflexes.

When the dominant is considered a necessary link of the conditioned reflex formation, there is a question whether the dominant and conditioned reflex have independent value in the organization of integral animal behavior in nature.

At different stages of development the dominant has different behavioral value. At the first stage, when the dominant focus is reinforced by the activity of the other centers, this is expressed as a dominant realization of a possible activity. Reinforcement of the defense reflex can lead to removal of the pain-inflicting stimulus and, hence, to the discontinuation of the dominant. It is the realization of inborn mechanisms, rather than learning, which, during the formation of a stable excitation focus, starts to sum up incoming stimuli, thus changing intercentral relations. At the stage of summation, the dominant has an independent value for the adaptive activity of the animal. The dominating activity is isolated.

At the stage of dominant specialization, when the dominant focus is reinforced by only one stimulus in the absence of responses to other stimuli, an element of learning is already present. When the discontinuation of the dominant is not achieved due to the activity of one reflex, excitation spreads to the other centers. The question as to how the motivation defense dominant is realized in animal behavior has already been considered in detail. Discontinuation of the motivation dominant also provides an element of learning realized at the expense of acquired experience or due to simulation.

When the dominant is realized as a "behavioral vector", a certain condition in the CNS acts as a trigger factor. The dominant condition arising both endogenously and exogenously initiates a behavioral reaction. In the classical conditioned reflex, a stimulus signalling forthcoming activity acts as a trigger initiating a behavioral reaction. Biologically, conditioned response is the preparation of the corresponding center for forthcoming activity directed at satisfying an apparent need, thus having great adaptive importance for animal behavior.

Attempts to divide the biological roles of the dominant and the conditioned reflex are quite schematic. Behavior as a whole is the realization of the system of temporary connections irrespective of whether it is triggered by a signal stimulus (in the presence of dominating need) or by the dominant state in the CNS. The dominant and the conditioned reflex belong to the common process of organization of goal-directed adaptive animal behavior.

REFERENCES

Asratyan, E.A. (1941) Principle of switching in conditioned activity. *Fiziol. Zh. SSSR*, **30**, 13−18 (in Russian)

Asratyan, E.A. (1953) *Fiziologiya Tsentral'noi Nervnoi Sistemy* (Physiology of the central nervous system). Moscow: USSR Academy of Medical Sciences (in Russian)

Asratyan, E.A. (1966) On functional architectonics of instrumental conditioned reflexes. *Zh. Vyssh. Nerv. Deyatel'nosti*, **16**, 577−588 (in Russian)

Asratyan, E.A. (1981) Bidirectional connection as a general neurophysiological principle. *Zh. Vyssh. Nerv. Deyatel'nosti*, **31**, 3−12 (in Russian)

Batuev, A.S. (1982) Principle of dominant in organization of goal-directed behavior. In *Teoreticheskie Voprosy Stroeniya i Deyatel'nosti Mozga* (Theoretical aspects of the brain structure and activity), vol. 12, pp. 69−80. Moscow: USSR Academy of Medical Sciences (in Russian)

Belenkov, N.Yu. (1979) Neuronal correlates of food behavior. In *Integrativnaya Deyatel'nost' Neironov* (Integral activity of neurons), pp. 81−82. Moscow: Meditsina (in Russian)

Beritov, I.S. (1932) *Individual'no-priobretennaya Deyatel'nost' Tsentral'noi Nervnoi Sistemy* (Individually acquired activity of the central nervous system). Tiflis: Gosizdat Gruzii (in Russian)

Drozdovska, G.Ya. (1987) Reciprocal inhibition in the case of blinking dominant in rabbits. *Zh. Vyssh. Nerv. Deyatel'nosti*, **37**, 439−447 (in Russian)

Frolov, A.A. and Sokolov, S.S. (1977) Reliability of estimates of the functions of spectral density and coherence using algorithm BPF. In *Metodika i Apparatura dlya Issledovaniya Psikhofiziologicheskikh Kharakteristik Cheloveka-operatora* (Methods and equipment for studying psychophysiological characteristics of man-operator), pp. 91−97. Moscow: Nauka (in Russian)

Hinde, R.A. (1970) *Animal Behavior.* New York: McGraw-Hill Book Co.

Konorski, J. (1967) *Integrative Activity of the Brain. An Interdisciplinary Approach.* Chicago: Chicago Univ. Press

Kupalov, I.S. (1949) On mechanisms of conditioned excitation. *Fiziol. Zh. SSSR,* **35,** 582–586 (in Russian)

Livanov, M.N. and Polyakov, K.M. (1945) Electrical processes in cerebral cortex of rabbits during elaboration of conditioned defense reflex to a rhythmic stimulus. *Izvestiya Akad. Nauk SSSR. Ser. Biol.,* **3,** 286–305 (in Russian)

Pavlov, I.P. (1949) *Pavlovskie Sredy* (Pavlov's wednesdays), vol. 2. Moscow-Leningrad: Izd-vo Akad. Nauk SSSR (in Russian)

Pavlov, I.P. (1951) Physiological mechanism of the so-called voluntary movements. In *Dvadtsatiletnii Opyt Ob'ektivnogo Izucheniya Vysshei Nervnoi Deyatel'nosti* (Twenty years experience of objective studies of the higher nervous activity), pp. 445–448. Moscow: Medgiz (in Russian)

Pavlygina, R.A. (1956) Formation of a dominant focus in hypothalamic area and a study of its properties. *Trudy Instituta Vysshei Nervnoi Deyatel'nosti,* **2,** 124–138 (in Russian)

Pavlygina, R.A. (1973) Dominant and conditioned reflex at the stage of generalization. *Zh. Vyssh. Nerv. Deyatel'nosti,* **23,** 687–695 (in Russian)

Pavlygina, R.A. (1979) Dominant as one of factors of behavior. In *Mekhanizmy Deyatel'nosti Mozga* (Mechanisms of brain activity), pp. 77–79. Moscow: USSR Academy of Sciences and USSR Academy of Medical Sciences (in Russian)

Pavlygina, R.A. (1980) Electrophysiological correlates of the motivation reaction in the case of dominant. In *Materialy 8-oi Vsesoyuznoi Konferentsii po Elektrofisiologii TsNS* (Materials of the 8th national conference on electrophysiology of the CNS), pp. 390–391. Erevan: Izd-vo Akad. Nauk Arm. SSR (in Russian)

Pavlygina, R.A. (1982a) Dominant and its importance in animal behavior. *Uspekhi Fiziol. Nauk,* **13,** 31–47 (in Russian)

Pavlygina, R.A. (1982b) The dominant and its role in animal behavior. In *The Learning Brain,* pp. 145–167. Moscow: Mir Publishers

Pavlygina, R.A. (1985) Stage of dominant specialization and goal-directed behavior. *Zh. Vyssh. Nerv. Deyatel'nosti,* **35,** 611–624 (in Russian)

Pavlygina, R.A. (1987) Role of dominant in the conditioned reflex formation. In *Dominanta i Uslovnyi Refleks* (Dominant and conditioned reflex). Moscow: Nauka (in Russian)

Pavlygina, R.A. and Lyubimova, Yu.A. (1987) Conditioned endogenous food reflex. In *XV S'ezd Vsesoyuznogo Fiziologicheskogo Obshchestva* (XV congress of the all-union physiological society), pp. 266–267. Leningrad: Nauka (in Russian)

Pavlygina, R.A. and Malikova, A.K. (1978) A study of defense dominant by using unconditioned blinking reflex. *Zh. Vyssh. Nerv. Deyatel'nosti,* **28,** 998–1003 (in Russian)

Pavlygina, R.A. and Malikova, A.K. (1985) Backward connections of the dominant and their role in goal-directed behavior. *Physiologia Bohemoslovaca,* **34,** 121–125

Pavlygina, R.A. and Sokolov, S.S. (1983a) The role of dominant in goal-directed animal behavior. In *Elektrofiziologicheskoe Issledovanie Stat-sionarnoi Aktivnosti v Golovnom Mozge* (Electrophysiological studies of steady state activity in the brain), pp. 34–54. Moscow: Nauka (in Russian)

Pavlygina, R.A., Malikova, A.K. and Lebedeva, M.A. (1983b) An electro-physiological study of the swallowing dominant. *Zh. Vyssh. Nerv. Deyatel'nosti*, **33**, 397–406 (in Russian)

Pavlygina, R.A., Rusinov, V.S. and Sokolov, S.S. (1983c) Electrical activity of the brain during behavior directed at the dominant discontinuation. *Zh. Vyssh. Nerv. Deyatel'nosti*, **33**, 344–355 (in Russian)

Roshchina, G.Ya. (1988) Influence of the dominant focus formed in the blinking center of the rabbit CNS in response to the limb defense reflex. *Zh. Vyssh. Nerv. Deyatel'nosti*, **38**, 72–80 (in Russian)

Rusinov, V.S. (1957) Electrophysiological studies of higher nervous activity. *Zh. Vyssh. Nerv. Deyatel'nosti*, **6**, 855–865 (in Russian)

Rusinov, V.S. (1972) Some new features of simple forms of the temporary relationship according to electrophysiological analysis. In *Elektriche-skaya Aktivnost' Golovnogo Mozga pri Obrazovanii Prostykh Form Vremennoi Svyazi* (Electrical activity of the brain during formation of simple temporary connections), pp. 3–22. Moscow: Nauka (in Russian)

Sawyer, C.H., Evereff, J.U. and Green, J.D. (1954) The rabbit diencep-halon in stereotaxic coordinates. *J. Comp. Neurol.*, **101**, 801–824

Shnirman, A.L. (1926) Combinative reflex and the dominant. In *Novoe v Refleksakh i Fiziologii Nervnoi Sistemy* (News about reflexes and phy-siology of the nervous system), pp. 2–16. Leningrad: Gos. Psikho-Nevrol. Akad. Press (in Russian)

Simonov, P.V. (1979) Emotions, memory and dominant. In *Gagrskie Besedy. Neirofiziologicheskie Osnovy Pamyati. Materialy 7-oi Konferentsii* (Gagra talks. Neurophysiological foundations of memory. Materials of the 7th conference), pp. 358–377. Tbilisi: Metsniereba (in Russian)

Simonov, P.V. (1983) Interaction between the dominant and conditioned reflex as a functional unit of behavior organization. *Uspekhi Fiziol. Nauk*, **14**, 14–23 (in Russian)

Skinner, B.F. (1953) *Science and Human Behavior.* New York: Appleton-Century

Skipin, G.V. (1951) On the mechanism of formation of conditioned food reflexes. *Zh. Vyssh. Nerv. Deyatel'nosti*, **1**, 922–926 (in Russian)

Struchkov, M.I. (1973) Forward and backward connections. *Uspekhi Fiziol. Nauk*, **4**, 26–41 (in Russian)

Sudakov, K.V. (ed.) (1979) *Sistemnye Mekhanizmy Motivatsii* (Systemic mechanisms of motivation). Moscow: Meditsina (in Russian)

Ukhtomskii, A.A. (1936) In *XV Mezhdunarodnyi Kongress Fiziologov* (XV International Congress of Physiologists), p. 62. Moscow-Leningrad: Izd-vo Akad. Nauk SSSR (in Russian)

Ukhtomskii, A.A. (1950) *Sobranie Sochinenii* (Collected works), vol. 1. Leningrad: LGU Press (in Russian)

Ukhtomskii, A.A. (1966) *Dominanta* (Dominant). Moscow-Leningrad: Nauka (in Russian)

Vasil'eva, V.M., Nezlina, N.I. and Ivannikova, T.N. (1966) Interaction between the swallowing reflex and artificial excitation focus in the rabbit motor cortex. *Zh. Vyssh. Nerv. Deyatel'nosti,* **16,** 1119–1120 (in Russian)

Vvedenskii, N.E. (1951a) On respiration in the frog. In *Polnoe Sobranie Sochinenii* (Full collected works), vol. 1, pp. 126–144. Leningrad: LGU Press (in Russian)

Vvedenskii, N.E. (1951b) Excitation and inhibition in reflex system during strychnine poisoning. In *Polnoe Sobranie Sochinenii* (Full collected works), vol. 4, pp. 202–269. Leningrad: LGU Press (in Russian)

INDEX

Conditioned reflex 61, 63, 65

Discontinuation of the dominant
52, 53, 57, 58, 60, 63, 64
Dominant 39–42, 51
—and dominant focus 39, 40, 43,
45, 46, 48, 54

Forward and backward
connections 44, 45, 47, 50, 54,
57

Goal-directed behavior 40, 42, 52,
53, 57, 59, 60

Hunger dominant 57–59

Motivation dominant 53, 55,
57–59, 62

Polarization dominant 53, 54, 60

Reciprocal inhibition 46, 51, 52

Summation 40, 45, 48, 52–54, 56,
58, 59

Swallowing dominant 42, 45, 48,
50

Temporary connection 42, 50, 51,
53, 54, 57–59, 62, 63
Thirst dominant 55, 56

Soviet Scientific Reviews, Section F
Physiology and General Biology Reviews

Contents, Volume 1 Part A

THE PROBLEM OF TIME IN DEVELOPMENTAL BIOLOGY: ITS STUDY BY THE USE OF RELATIVE CHARACTERISTICS OF DEVELOPMENT DURATION T.A. Dettlaf, G.M. Ignatieva and S.G. Vassetzky

DIFFERENTIATION OF HEMOPOIETIC CELLS IN DIFFERENT EXPERIMENTAL SYSTEMS N.G. Khrushchov and T.V. Michurina

ACETYLCHOLINE AND BIOGENIC MONOAMINES AS INTRACELLULAR REGULATORS OF EARLY EMBRYOGENESIS G.A. Buznikov

FORMATION OF THE ENZYMIC APPARATUS DURING DIFFERENTIATION L.S. Milman and Yu.G. Yurowitsky

INVESTIGATIONS OF REGULATORY MECHANISMS IN FISH EMBRYOS: AMOUNT OF MITOCHONDRIA, LOW MOLECULAR WEIGHT RNAs AND TEMPERATURE-DEPENDENT POLYAMINE SYNTHESIS A.A. Neyfakh, N.B. Abramova, and T.A. Burakova

BIOCHEMISTRY OF EMBRYONIC INDUCTION: IDENTIFICATION AND CHARACTERIZATION OF MORPHOGENETIC FACTORS A.T. Mikhailov and N.A. Gorgolyuk

EGG CORTICAL REACTION DURING FERTILIZATION AND ITS ROLE IN BLOCK TO POLYSPERMY A.S. Ginsburg

INTERACTION OF T-LYMPHOCYTES WITH HEMOPOIETIC STEM CELLS: ITS INFLUENCES ON THE PROLIFERATION AND DIFFERENTIATION PROCESSES OF HEMOPOIETIC PRECURSOR-CELLS R.V. Petrov and V.M. Man'ko

Contents, Volume 1, Part B

SOME MOLECULAR-GENETIC ASPECTS OF CELLULAR DIFFERENTIATION IN *DROSOPHILA* L.I. Korochkin, M.Z. Ludwig, E.V. Poliakova and M.R. Philimova

DNA SUPERCOILING IN DIFFERENTIATION AND IN GENE SWITCHING IN EUCARYOTES A.N. Luchnik

DNA POLYMERASES IN THE EARLY DEVELOPMENT OF A TELEOST FISH *MISGURNUS FOSSILIS* (LOACH) V.S. Mikhailov

THE CLONING AND STRUCTURE OF GENES CODING FOR THE EYE LENS CRYSTALLINS OF THE FROG G.G. Gause and S.I. Tomarev

GENE EXPRESSION AND CELL INTERACTIONS IN THE MAMMALIAN DEVELOPMENT B.V. Konyukhov

EUCARYOTIC MOVABLE ELEMENTS G.P. Georgiev

STUDIES OF CELLULAR, CHROMOSOMAL AND MOLECULAR MECHANISMS OF VERY EARLY MAMMALIAN EMBRYOGENESIS A.P. Dyban

TRANSPLANTATION OF NUCELI AND GENES INTO ANIMAL OVA K.G. Gazaryan

Contents, Volume 2 Part A

MECHANISMS OF TRANSMISSION IN THE INTERNEURONAL SYNAPSES OF VERTEBRATES A.I. Shapovalov

MECHANISMS OF SYNAPTIC TRANSMISSION IN THE SYMPATHETIC GANGLION V.I. Skok

DESCENDING CONTROL OF THE ACTIVITY OF THE BRAIN STEM STRUCTURES V.V. Fanardjian

SYNAPTIC HIPPOCAMPAL ACTIVITY *IN VITRO* V.G. Skrebitskii

NEURONAL ORGANIZATION OF THE HYPOTHALMIC MECHANISMS OF REGULATION OF AUTONOMIC FUNCTIONS O.G. Baklavadjian

Contents, Volume 2, Part B

NEUROGLIA: PROPERTIES, FUNCTIONS AND SIGNIFICANCE IN NERVOUS ACTIVITY A.I. Roitbak

THE PHENOMENON OF SPATIAL SYNCHRONIZATION OF BRAIN POTENTIALS IN INVESTIGATIONS INTO THE SYSTEM ORGANIZATION OF THE ACTIVITY OF THE BRAIN M.N. Livanov and V.N. Dumenko

INTERRELATIONSHIPS OF BRAIN HEMISPHERES V.M. Mosidze

FUNCTIONAL CHARACTERISTICS OF NERVOUS TISSUE (HIPPOCAMPUS AND SEPTUM) TRANSPLANTED INTO THE ANTERIOR EYE CHAMBER AND BRAIN O.S. Virogradova

INVESTIGATIONS OF NEUROPHYSIOLOGY AND MECHANICS OF PLASTICITY AND LEARNING N.L. Voronin

THE NEOSTRIATUM: NEUROPHYSIOLOGY AND BEHAVIOUR N.F. Suvorov, S.V. Al'Bertin and N.L. Voilokova

MONOAMINERGIC BRAIN SYSTEMS AND THEIR ROLE IN THE REGULATION OF BEHAVIOUR E.A. Gromova

Contents, Volume 3

Part 1

POPULATION ECOLOGY AND ECOPHYSIOLOGY I.A. Shilov

Part 2

POPULATION CYTOGENETICS OF ANIMALS V.N. Orlov and N.Sh. Bulatova

POPULATION RADIATION ECOLOGY OF ANIMALS D.A. Krivolutskii

Part 3

POPULATION BIOLOGY OF AMPHIBIANS V.G. Ishchenko

POPULATION BIOLOGY OF INVERTEBRATES S.O. Sergievskii

Part 4

ANIMAL POPULATION GENETICS L.Z. Kaidanov

POPULATION ETHOLOGY E.N. Panov

Part 5

POPULATION PRINCIPLES IN RESEARCH INTO NATURAL FOCALITY OF ZOONOSES E.I. Korenberg

PROGRESS AND PROBLEMS IN POPULATION BIOLOGY A.V. Yablokov

Contents, Volume 4

Part 1

POPULATION PROBLEMS IN THE BIOLOGY OF UNICELLULAR ORGANISMS I.I. Gitel'zon, N.S. Pechurkin and A.V. Bril'kov

Part 2   Population Biology of Phytopathogenic Fungi and Plants

POPULATION BIOLOGY OF PHYTOPATHOGENIC FUNGI Yu. T. D'yakov and M.V. Godenko

POPULATION BIOLOGY OF PLANTS M.M. Magomedmirzaev

Part 3

PERSPECTIVES OF POPULATION PHENOGENETICS V.M. Zakharov

Part 4

NEURONAL BASIS OF ASSOCIATIVE LEARNING V.M. Storozhuk

NEURONAL MECHANISMS OF SHORT-TERM MEMORY A.S. Batuev

Part 5   The Problems of Higher Nervous Activity

THE NEUROPHYSIOLOGY OF THE EMOTIONS P.V. Simonov

NEUROPHYSIOLOGY OF THE EMOTIONS AND THOUGHT N.P. Bechtereva, Yu. L., Gogolitsyn and D.K. Kambarova

Contents, Volume 5

Part 1

PROBLEMS OF INTERNAL INHIBITION U.G. Gassanov

THE DOMINANT AND THE CONDITIONED REFLEX R.A. Pavlygina

Part 2

RELAY FUNCTIONS OF HIPPOCAMPAL MONOAMINES IN ACQUIRED AND INBORN FORMS OF BEHAVIOR G.G. Gasanov and E.M. Melikhov

PHYSIOLOGICAL MECHANISMS OF COMPLEX BEHAVIOR IN ANTHROPOIDS L.A. Firsov, V.A. Syrenskii and T.G. Kuzretsova

Part 3

A NEW IDEOLOGY OF STUDIES OF THE NEUROPHYSIOLOGICAL CORRELATES OF MENTAL ACTIVITY S.V. Medvedev and S.V. Pakhomov

Part 4

THE NEURAL MECHANISMS OF CONSCIOUS AND UNCONSCIOUS PERCEPTION E.A. Kostandov